MOUNTAINS BEFORE MOUNTAINEERING

Praise for Mountains Before Mountaineering

Engrossing, astonishing, thought-provoking; shatters the long-standing illusion that mountains were held in abhorrence during the early modern era. Fascinating to read about mountains as seen through the eyes of early modern travellers, writers, poets, philosophers, naturalists, artists, map-makers, and the people who actually lived there.

Jo Woolf, author of *Britain's Landmarks and Legends*

Like a passionate guide, Dawn Hollis leads you along the little-travelled paths of pre-modern mountains ... With a sharp and reflexive eye for the practice of historical research, she challenges the prevailing notion of 'Mountain Gloom' associated with that era. Spanning centuries and continents, she unveils the diverse perceptions and uses of mountains beyond the single practice of mountaineering.

Dr Gilles Rudaz, University of Geneva

Dawn Hollis's rich and compelling account will transform the way we think about the history of human engagement with mountains ... Based on painstaking research across a vast range of sources, it brings to life the responses of individuals and communities whose stories have been sidelined from traditional histories of mountains and mountaineering.

Professor Jason König, University of St Andrews

Dawn L. Hollis's thorough research strips the myth down to uncover our longer human respect, curiosity, and affection for the mountains that predates mountaineering. Hollis' *Mountains before Mountaineering* challenges us to reconsider our human relationship with mountains and who we are as adventurers and people.

Andrew Szalay, *The Suburban Mountaineer*

Mountains before Mountaineering destroys the myth that mountaineering is a quintessentially modern pursuit, or that pre modern people feared mountains ... Hollis is well aware of the exploitation of local communities and ecologies that mountaineering now so often entails. Packed with vivid anecdotes, this book is a historical anthropology of the peak.

Lyndal Roper, Regius Professor of History, University of Oxford

MOUNTAINS BEFORE MOUNTAINEERING

THE CALL OF THE PEAKS BEFORE THE MODERN AGE

DAWN L. HOLLIS

First published 2024

The History Press
97 St George's Place, Cheltenham,
Gloucestershire, GL50 3QB
www.thehistorypress.co.uk

© Dawn L. Hollis, 2024

The right of Dawn L. Hollis to be identified as the Author of this work has been asserted in accordance with the Copyright, Designs and Patents Act 1988.

All rights reserved. No part of this book may be reprinted or reproduced or utilised in any form or by any electronic, mechanical or other means, now known or hereafter invented, including photocopying and recording, or in any information storage or retrieval system, without the permission in writing from the Publishers.

British Library Cataloguing in Publication Data.
A catalogue record for this book is available from the British Library.

ISBN 978 1 80399 318 8

Typesetting and origination by The History Press
Printed and bound in Great Britain by TJ Books Limited, Padstow, Cornwall.

 Trees for Life

Contents

Introduction	7
A Note on Transcriptions	21
Vantage Point: Climbing Wine Barrels, Climbing Alps	23
Chapter One: Mountain Ventures and Adventures	31
Vantage Point: A Chapel in the Mountains	61
Chapter Two: The Real Mountaineers	65
Vantage Point: Healing the Blind	97
Chapter Three: The Meanings of Mountains	101
Vantage Point: Into the Volcano	139
Chapter Four: Mysteries of Science, Mysteries of Faith	143
Vantage Point: But Who Was First?	173
Chapter Five: How a Myth Becomes History	175
Epilogue: Mountains After Mountaineering	203

Notes	211
Bibliography	225
Acknowledgements	233
Index	237

Introduction

In 1786, a doctor and a peasant burst into a spontaneous race for the final few metres of their shared journey to the summit of Mont Blanc, which had never before been conquered. Over the following century, the rest of the Alps became the 'playground of Europe', as triumphant first ascents were made across the range. In 1953, a New Zealand beekeeper and a Nepali-Indian Sherpa claimed Mount Everest for Britain; headlines celebrated, on the day of Queen Elizabeth's coronation, 'the crowning glory' of Hillary and Tenzing reaching the summit of the world. Today, the mountains of the world attract innumerable visitors, ascending them in hiking boots and crampons, or descending them with adrenaline-filled rapidity upon skis, snowboards, or rugged bicycles. They are photographed, painted, and admired. This era of modernity – the era we still inhabit – is one in which mountains are places of heroism, of joyous sporting endeavours, of beauty and sublimity. But what came before?

Before this time, or so it is said, travellers shuddered at the very sight of the Alps. Peasants who lived in the shadows of the mountains whispered that the summits were the abode of dragons, best avoided. Mountains were ugly, seen as warts upon the face of the Earth. The very thought of climbing to the top of a mountain was an absurdity. It was only in the modern era, with the start of mountaineering and the new appreciation for nature, as expressed in the writings of poets such as William Wordsworth, that mountains came to be objects of love and admiration. One might say that today we are the inheritors of a uniquely modern appreciation for the natural landscape.

This is a compelling vision. It is fascinatingly strange to imagine a time when the overall view of mountains was so at odds with that of today, and deliciously tempting to cast ourselves as having a special relationship with mountains, unknown by our ancestors. It's a vision you might be familiar with, since it is embedded in accounts of the origins of mountaineering, in articles about aesthetics and art, and in many ways in the very fabric of what it means to 'be modern'. It is also, as I have discovered, *wrong*.

I am at my dad's retirement party, on a visit back to the parental nest long after flying it for marriage and PhD research up in Scotland. For thirty-five years he worked for the local district council, in building control and planning permission. I am surrounded by Suffolk accents and builder-y types, who are about as far removed from my daily work of poring over centuries-old books as I am from inspecting walls for safety or architectural sketches for planning violations. One of my dad's colleagues, who I am pretty sure dandled me on his knee when I was a toddler, pauses to ask me about my doctorate. What do I work on? The conversation goes a bit like this:

'Oh, er, mountains. In history.'

'Mountains, really? What about them?'

'Well, I'm looking at mountains in the seventeenth century, what people thought about them ...'

'Oh, yes! People didn't like mountains back then, did they?'

'Well, actually ...'

The same conversation plays out in any number of settings: making small talk at the doctor's; at academic conferences; sitting next to random people on the bus. I am always amazed at the fact that this idea, that *people didn't like mountains then*, seems to have burrowed into the collective unconscious. Early on in my PhD, I read a book by the historian Daniel Lord Smail which talks about 'ghost theories'.[1] I like this term, because it seems to capture exactly what I keep coming up against: an idea so old and so oft-repeated that it has taken on the status of fact, its real origins long forgotten.

I could share a dozen examples of this theory as expressed on television, in books, or in magazine articles. My favourite is an old one, from Kenneth

Clark's 1969 *Civilisation*, a then-groundbreaking documentary on the history of art, which was intended partly to take advantage of the full potential of the new technology of colour television. Seated on a large rock on a mountainside, incongruously dressed in a suit and tie, he declared in clipped, definite accents that:

> For over two thousand years mountains have been considered simply a nuisance: unproductive; obstacles to communication; the refuge of bandits and heretics. It's true that in about 1340 the poet Petrarch had climbed one, and enjoyed the view at the top ... and at the beginning of the sixteenth century Leonardo da Vinci had wandered about in the Alps ... No other mountain climbs are recorded.

He went on to observe that to most people 'the thought of climbing a mountain for pleasure would have seemed ridiculous', and when 'an ordinary traveller of the sixteenth and seventeenth centuries crossed the Alps, it never occurred to him to admire the scenery'.[2]

This book tells a different story. Long before Everest was 'discovered' as the highest mountain in the world, long before the first (recorded) ascent of Mont Blanc, mountains in fact inspired curiosity and fascination. A wonderful summation of this can be found in an oration on travel written by Hermann Kirchner (1562–1620), a professor of history and poetry at the University of Marburg. His goal was to urge young men to journey into foreign and distant lands for their edification and personal improvement. Mountains offered a special attraction:

> What I pray you, is more pleasant, more delectable, and more acceptable unto a man than to behold the height of the hills ... to view the hill Olympus, the seat of Jupiter? to pass over the Alps that were broken by Hannibal's vinegar? ... to behold the rising of the Sun before the Sun appears? to visit Parnassus and Helicon, the most celebrated seats of the Muses?[3]

Professor Kirchner invited his readers on a journey through the mountains. This book does the same, inviting you on a journey through mountains as they were viewed, experienced and loved before the modern age.

My Journey to the Mountains

I have now been studying the history of mountains for most of my adult life, long enough that I have replaced the verb of studying with that of being: I *am* a historian of mountains, specifically of early modernity. I am going to talk more later about that term, but for the time being let's just say that by 'early modern' I mean the period between roughly 1450 and 1750. Before drilling down into questions of terminology, though, I want to explain how I got here. What journey did I take to becoming a historian of mountains, and to doubting the ghost theory that people didn't like mountains back then?

There are many places this story could begin, because it starts not with history but with my own personal relationship with mountains. Maybe it starts with childhood holidays and hiking up hills in the Lake District, the Yorkshire Dales, Northumbria. I particularly loved, and still love, the Lake District, the names of the peaks rolling off my tongue and around my head like incantations: Blencathra, Scafell, Skiddaw, Helvellyn. These names evoke visceral teenage memories: getting soaked and cold slipping up the shaley summit of Skiddaw, the foot-bruising monotony of the upper moonscape of Scafell Pike, the heady thrill of tracing the narrow stony top of Striding Edge up to the summit of Helvellyn.

Or maybe it starts with that peculiarly masochistic form of British youth improvement (I wonder what Hermann Kirchner would have thought of that), the Duke of Edinburgh award, and the two 'Gold' expeditions trekking through wild country with a tent and what felt like the kitchen sink on my back. At one memorable point on the second of the two expeditions, the team member, whose turn it was to route-find, stopped us at the top of a hill, glancing with dawning horror between the map and the peak across from us, before uttering the sentence: 'I think we climbed the wrong mountain.' Somehow, this still did not put me off.

I was certainly well down the path to my mountains of today – or maybe I should say my mountains of yesteryear – when the teacher who ran the Duke of Edinburgh scheme at my school lent me a pile of books on the Everest expeditions of the 1920s, and the 'mystery of Mallory and Irvine'.[4] George Mallory and Sandy Irvine, members of the 1924 team hoping to claim Everest for Britain, departed their tent for their summit bid on the morning of 8 June 1924. Later that afternoon, cloud enveloped the top of the mountain, and they were never seen alive again. The 'mystery' lay in the tantalising suggestion that they might have successfully reached the 'top of the world' almost thirty years before Edmund Hillary and Tenzing Norgay.

The same teacher also led an annual winter mountaineering weekend to Scotland, during which I wore crampons for the first time and learned to self-arrest with an ice axe. Two years later, she took a sabbatical from work in order to make her own attempt upon Mount Everest. My peers and I at school waited for updates coming down the mountain from her expedition team with bated breath. Some of us had figured out, with the obsessiveness of teenagers towards a beloved role model, that if she succeeded she would break the record for being the oldest British woman to summit Everest. This was apparently news to her; when she returned, she described with some indignation the experience of returning to Base Camp from her successful summit attempt to be vaunted for her supernumerary record-breaking. I dreamed of following in the footsteps both of my teacher and of Mallory and Irvine. I signed up for a week-long winter mountaineering course in the Highlands, and climbed Ben Nevis, Buachaille Etive Mòr, Stob Coire nan Lochan – more names to conjure with – with the North Face of Everest rising in my mind's eye.

For me, though, it has always been peaks and books, the hiking boots and the armchair. As I worked on my fitness in the school gym in the mornings, in the evenings I filled my mind with every scrap of information I possibly could about the first three Everest expeditions of 1921, 1922 and 1924: George Mallory's expeditions. It was like a historian's Rubik's cube. I wondered whether, if I turned the evidence over in my mind enough, I could fit it all together in just the right way to prove that they had made it.

Around this time, I applied to Oxford, another summit that rose high in my imagination, and was invited to interview for my chosen subject of history. Now, Sandy Irvine – Mallory's climbing companion – had been a student at Merton College, Oxford, and his letters from the expedition were in the library there. So I wrote to the archivist of Merton College. I was coming up to Oxford, I said (and it is always *up*, like a mountain, that one goes to Oxford, no matter where in the world you are travelling from), and was wondering whether I could come look at the letters, diaries and photos they held from the 1924 expedition. To my amazement they said yes to my juvenile enthusiasm, and in between nerve-wracking interviews which I was convinced I had bungled, I made my first ever research trip to an archive, pencil and notebook in hand, ready to meet Sandy Irvine.

Fifteen years later, I can still remember how utterly electrifying it felt to sit in that leather chair, in that wood-panelled library *in which my research subject would have studied*, touching letters he had touched, my fingers tracing his handwriting, reading the words he had written to his mother. With the damp winter darkness outside and the warm lamplight inside, it was like the memory of a childhood Christmas; an inexpressible treat in the midst of winter. Sandy Irvine was 22 years old when he died, an engineering student, a Blues rower: tall, broad-shouldered blond. He was the 'experiment' of the 1924 expedition, a young man with immense strength but relatively little mountaineering experience. Over the course of the trek into Tibet, he became self-appointed tinkerer and technical expert, fixing watches and more or less redesigning the apparatus for inhaling bottled oxygen at high altitude (itself another experiment).[5] He was also, I discovered that December, funny and good company.

It might seem strange to speak of someone who is long dead as 'good company', but that was truly how I felt reading Sandy Irvine's irrepressible letters to his mother. Years later, a conversation with an eccentric tutor on my Master's degree showed me I was not alone in feeling this way. I was writing my dissertation on Thomas Burnet (c.1635–1715), who you will meet later, and fell into conversation with the tutor after a session on book history. 'I hear,' he said, rocking backwards and forwards on his heels, 'that you are working on one of my friends.'

I frowned as I tried to parse this sentence. Working 'with' one of his friends might have made sense, if that was how he identified the professor supervising my dissertation. But 'on'? His impatient clarification cleared this up. 'Thomas Burnet, of course! You know, I have a copy of his *Theory of the Earth* …'

Since then, I have rather enjoyed the idea of identifying the historical figures one studies as friends, acquaintances, or even (perhaps) enemies. I like people and I like uncovering the traces they leave of themselves, even in supposedly impersonal theological writings such as those composed by Burnet in the late seventeenth century. Throughout this book I will be introducing you to my friends, and making sure their mountain stories do not vanish into historical obscurity like Mallory and Irvine did into the fog of Everest.

Introduction

❧ ❧

Sandy Irvine made me giggle in the archive from his very first letters home, on the boat out to India. A week into the voyage, he commented drily that 'One passenger only so far has died, and I didn't know in time to see the funeral which was a disappointment.'[6] Trekking through Tibet, he discovered that his assumption that the local inhabitants would not mind nudity was quite mistaken, and he had come without a bathing gown. Fortunately, he managed to cobble one together 'out of a belt and two handkerchiefs' – the mind boggles as to what this looked like.[7] He did not always jest, however. The use of oxygen at high altitude was a new and controversial technology, with many climbers worrying that it represented using 'unfair means' against the mountain. Despite being the oxygen mechanic, Sandy was also conflicted about it, concluding seriously that 'I think I'd sooner get to the foot of the final pyramid without oxygen than to the top with it'.[8] He was also intrigued by the foreign land and culture he found himself immersed in, and sent descriptions and sketches of Tibetan life back to his family in Cheshire, along with occasional commentary on the 'charming' appearance of the local girls.

One line of discussion within the wider mystery of Mallory and Irvine is precisely why George Mallory chose the inexperienced youth for his climbing partner on that fateful summit attempt, and not one of the many other more seasoned mountaineers on the expedition. Whether it was for his strength (Irvine consistently outperformed the older men on their regular high-altitude fitness tests) or his technical expertise with the oxygen equipment that Mallory elected to use for the summit bid, the case remains that this choice tied their names together in the history books. George Mallory had 37 years to Sandy's 22; dark hair beside his blond; slight, long-armed height to Sandy's Odyssean broad chest. He was a light-blue Cambridge man, had studied history, and was a writer of long, eloquent passages rather than dashed-off, joking letters. He also had a penchant for being photographed naked – no handmade bathing gowns for him.

By the time I finally came to write my undergraduate dissertation – at Oxford, which I got into despite my pessimism at interviews – my interest in these two men had developed beyond the simple question of whether or not they summited Everest. Inspired by Irvine's notes on the ways (and women) of Tibet, I decided to write about the moments of encounter between the British members of the 1920s expeditions and the Tibetan people whose villages they travelled through and whose labour they relied upon to haul tents and other supplies up the mountain.

※ ※

Over my three years at Oxford, two important things occurred: I continued to climb mountains and I decided that I wanted to become an early modern historian. I joined the university mountaineering club, which, despite the name, was mostly focused on rock climbing. Towards the end of my first year in the club, a novice climber started flirting with me, and I unwisely attempted to teach him how to belay on an indoor climbing wall. I fell and was only caught a few feet above the floor when another club member grabbed the end of the rope that my Casanova should have put a brake on. Not long after, I met someone who preferred hillwalking to roped climbing anyway, and we became engaged on the summit of Blencathra in the Lake District.

Meanwhile, despite my fascination with twentieth-century mountaineering, I found myself increasingly drawn towards the early modern period of history: the era which slots between the end of the medieval age and the Industrial Revolution. For many people, the term 'early modernity' signals a sense of familiarity, the idea that it was the period in which the modern world as we know it was starting to begin. To me, early modernity is actually like that moment of falling: a sense of suspended exhilaration, as that which is certain falls away, and the historian travelling progressively backwards in time begins to encounter people whose modes of thinking are entirely distinctive from those of today. By the end of my undergraduate degree I knew this was the period I wanted to study in my Master's and PhD.

It took a critical moment of route-finding to bring my two historical obsessions – mountains and early modernity – together. I was lucky in my choice of guide: the tutor with whom I had taken a first-year course on the early modern period. Brilliant and self-effacing, she exploded my preconceptions of what an Oxford don could be. Three weeks into the course, she had glanced down at the reading list we had been given and tapped her finger on the title of a book she herself had written. 'You know,' she mused, 'I'm not entirely sure the author was right in her conclusions. What do you all think?' She was also generous: when I told her, two years later, that I was thinking of applying to do postgraduate research, she offered to help me with my proposals.

I had a long list of different topics that I might write a Master's dissertation and, ultimately, a doctoral thesis on, which I handed to her. Early modern experiences of pregnancy? Mental illness, with a focus on the early modern concept of 'melancholy'? She shook her head at these as too well researched already. (For historians, as well as mountaineers, the less trodden ground is often the most attractive.) At the very bottom of the list I had written 'mountains'. I had left them till last because it had seemed to me to be almost too greedy to imagine that I might get away with combining my passion with my research. My mentor stabbed her finger at the word. 'Here you go,' she said. This was the topic I should do – mountains were 'cool', and not a lot had been written about them in the early modern period since a book published in 1959: *Mountain Gloom and Mountain Glory*, written by Marjorie Hope Nicolson, the first woman to hold a full professorship at the University of Columbia.

As any mountaineer knows, once you have looked up at the route ahead of you it can be hard to turn back. That moment led me, eventually, to sitting in my home office in the midst of the COVID-19 pandemic, writing the first draft of this book.

Mapping this book

Like a map, a book has limits. It will cover an area in detail, but there will always be blank space off the edges. To define the scope of a book of history you must answer four questions: what, when, where and how.

What is this book about? Mountains. But what exactly is a mountain? The answer to this question is more complicated than you might expect. Fairly early on in my research into early modern mountains, I gave a presentation at a conference intended for postgraduate students to gain their first experience of public academic speaking. This was back in the days when smartphones were still a relative novelty, and the act of quickly googling something mid-conversation had yet to become a widely accepted habit. During questions, one of the 'real academics' in the audience asked me what I meant by a mountain. I can't quite remember what I answered – I think I said it was a pretty vague term – but I do remember that, just after I had answered the question following his, he looked up from his phone and interrupted triumphantly, 'It's 2,000ft! A mountain is 2,000ft! Anything lower and it's a hill.'

This reminds me of the now disconcertingly vintage film, *The Englishman who Went up a Hill but Came down a Mountain*. (In my PhD thesis, I referred to this film as featuring a 'fresh-faced Hugh Grant', and my supervisor, who I think deemed himself a contemporary of Grant, commented with a touch of desperation, 'isn't he still fresh-faced?') The conceit of the film, set in 1917, is that two English cartographers arrive at a Welsh village to survey the local 'mountain', only to discover that it measures just a little short of the required 1,000 (in this case) feet. The community respond by carrying earth up to the top of the hill/mountain to build the summit up to the required height. There is also the inevitable love interest for Hugh Grant and a certain quantity of mountain/hilltop embraces.

The thing is that my seventeenth-century friends never watched Hugh Grant pretending to be a cartographer, and lived long before the US Board on Geographic Names defined a mountain as 1,000ft, or before the UK government placed the line at 2,000ft, or before Sir Hugh Munro's list of Scottish peaks gave special prominence to summits over 3,000ft. They also lived before the United States Geological Survey concluded that there is no technical definition of a hill versus a mountain.[9] As you will find as you read on, early modern landscape viewers used the terms hill and mountain with happy interchangeability, and landforms which we might today consider relatively small, on a global scale, could seem enormous to people who had never seen the Alps or even heard of the Himalayas.

It is necessary, then, to leave behind modern categories of scale. At the risk of being just as vague as I was at that conference, I am interested in what people thought about any hilly or mountainous landscape, no matter how insignificant the elevation might seem to a modern mindset that thinks of mountains in terms of high, higher and highest. Before modernity, what made a mountain was in the eye of the beholder.

In terms of the 'when', it should already be clear that this book focuses on the early modern period. The term is one which historians bandy about quite happily, without realising that it is actually somewhat obscure to people who are neither students nor researchers in the discipline. It doesn't much help that historians cannot precisely agree on the date range it describes. At its broadest extent, I would say it ranges from about 1450 at the earliest to 1800 at the latest, but I can't promise that you won't find an academic who would date it as starting in 1400 or ending in 1850.

My attention in this book will largely be on the sixteenth and seventeenth centuries (1500–1700). However, this book will share stories from both before and after those dates. Before, because no book on historical mountain experiences could go without reference to Petrarch's famous (and problematic) ascent of Mont Ventoux in 1336. After, because the development of modern mountaineering is central to explaining that conversation I had at my dad's retirement party, to understanding why the mountain stories I will be sharing herein have been so long neglected and forgotten.

Throughout this book I will sometimes refer to the 'premodern period', or to 'premodern experiences of mountains'. The traditional history of mountain attitudes tends to tar all of the eras which came before the modern period with the same brush: today, we love mountains, but back then, whether early modern, medieval, or classical, they did not. I think that story is wrong about the early modern period and is wrong about ancient times too. Although it does not form the main focus of this book, there is a wealth of material and scholarly work out there that reveals rich and complicated relationships between people and mountains across different cultures and time periods.[10] Moreover, early modern responses to mountains were influenced by, and echoed, older experiences and traditions. It is this sense of connection and continuity which I intend to flag by using the term 'premodern' as well as 'early modern'.

To return to the mapping metaphor, the book operates at three different scales of time and detail. At the widest scale, this book will tell you about mountains before modernity, and will also give a sense of what changed (and what did not) during modern times. At the intermediate scale, it offers a slightly closer view of mountain responses from the Renaissance to Romanticism. At the smallest scale, the highest density of detail will provide you with a clear sense of the contours of mountain attitudes and ideas during the two centuries at the heart of the early modern period.

There are many mountains in the world. Which mountains, where, am I interested in? Yet again there is no tidy answer. This book will focus upon the engagements that Europeans had with mountains. Even in the sixteenth and seventeenth centuries, however, travellers ranged far afield, and geographically speaking the mountains in this book include the Elburz range and the hills of the Holy Land.

At another talk about my work, another well-meaning audience member asked me why I had not included anything about mountains in ancient Chinese literature. I am sure there is material out there. I am equally sure that a global history of premodern mountains would fill far more than a single volume. Early modern Europe, on the other hand, represents a more manageable scope. It is also, due to the particularly cultural flows of the period, a category which makes coherent sense. Europe in this period was subject to fairly self-contained currents of *ideas* – books, published in the common learned language of Latin, made their way from the book fairs of Frankfurt to the private libraries of Britain; British natural philosophers (scientists) were published in Amsterdam, and so on. There was a shared European culture in this period which underpinned responses to mountains whether in Wales or Westphalia.

Finally, we reach the 'how', by which I mean what sources, whose stories am I using to fill in my map of early modern mountains in the European mind? With a sigh, I must respond: mostly dead white men. They were also men who, whilst they might not all have thought of themselves as wealthy, were highly privileged in terms of their access to education. These were the individuals who possessed the literacy necessary to record their thoughts about mountains. They were also privileged in that their words have been preserved for us to read today, largely thanks to having been published. (More copies means more chances for at least one copy to survive.) Many people in early modern Europe could not read or write, and relatively few of those who could would have had the means, status or connections to get their words into print.

That is not to say that poor people, or women, or members of minority cultures did not travel among, or have emotions evoked by, the mountains. They did, and traces of their experiences survive in archaeology, in ethnographic evidence, and indeed in the writings of those dead white men. Chapter 2 shines a spotlight on the 'real mountaineers' of early modern Europe: the people who lived and worked among the mountains. But confining them to

Introduction

one chapter in this way is not to imply anything about the relative importance of their experiences; it is merely a reflection of the relative wealth of sources surviving for, well, the wealthy and privileged.

⁂

It remains only for me to offer a brief route description to guide you through this book. It is divided into five chapters, each one preceded by a 'Vantage Point'. Each vantage point shares just one or two historical sources which encapsulate the theme of the chapter to follow. You could, if you just wanted to read a few brief stories that summarise the key themes of this book, just read the vantage points. The chapters include more stories, but they also go into more depth exploring some of the 'workings out' involved in the writing of history. Studying history is about a lot more than just figuring out 'what really happened'; it is about figuring out what 'truth' we can recover from biased historical sources, deciding what sort of information we *can* get from things like poems or works of art, and about recovering the experiences of people who did not get to record their own words and ideas for posterity. In this book, I hope to give you an insight into the challenges and debates surrounding all of these things.

The chapters will also contain a bit more about my own journey, as well as those of my early modern 'friends'. It is not, as one might expect of a mountain-themed book, a journey involving my own heroic escapades up snowy peaks. Instead, it is my journey of the mind, of moments of discovery and understanding which helped me cross the crevasses of time in order to begin to understand mountains as my early modern writers did. For me, that journey has been just as exciting and enriching as any made with crampons and ice axe.

Fittingly, then, the first viewpoint and chapter will start with journeys, tracing the stories of some hardy (and not-so-hardy) travellers to the mountains. The second chapter will focus not on travellers but on mountain-dwellers: the 'real mountaineers' who knew almost as much about mountain safety as climbers of today. In Chapter 3, I tackle possibly the most difficult sources of the whole book: artworks and poetry. These do not necessarily tell us much about what *happened* on mountains in the past, but they do help us figure out what they meant to people. Chapter 4 turns to the question of early modern mountain science. This was an era before modern geology and before an understanding of deep time, and writers interested in the history of the Earth got into some surprisingly heated arguments trying to figure out how mountains were formed. The final chapter then turns back to the problem this introduction started with: the idea that people 'didn't like mountains back then'. I explain where that idea came from, and what it tells us about our own, modern, relationship with mountains.

    I think that is all the map required to navigate this book. Let us begin the journey.

A Note on Transcriptions

Throughout this book I quote from texts written during the sixteenth and seventeenth centuries and earlier. Variations in spelling and printing conventions (the use of a long 'f' for an 's' in the middle of a word, or the use of a 'u' to represent a 'v') mean that writing from this period can appear somewhat obscure to modern eyes. Since my goal is to introduce you to my early modern friends and to help them share their stories and ideas about mountains, I have silently modernised spellings in all quotations, though not in the titles of books. The references provided should enable you to find the original words – unusual spellings and all – should you so wish.

Vantage Point: Climbing Wine Barrels, Climbing Alps

To get a sense of the irrepressible personality of Tom Coryate (1577–1617), traveller and courtier, you need look no further than his account of ascending not a mountain, but a wine barrel.

This wine barrel, located in the palace at Heidelberg, Germany, was 'the most remarkable and famous thing' that he saw on his extensive journeys. Had it been around in ancient times, Coryate believed it would have been added to the list of wonders alongside the Colossus of Rhodes and the Hanging Gardens of Babylon. Even the 'gravest and constantest man in the world' (a phrase which most certainly does not describe Coryate) would have been struck with wonder at the sight. What was so special about this wine barrel? Well, it was huge. Coryate's book included an illustration, which he claimed was made from accurate sketches that he took at the time, showing a barrel five times the height of the people gathered at its base. The woodcut also shows the man himself, balanced on top of the barrel 'with a cup of Rhenish wine' in hand. Coryate described his ascent, up twenty-seven rungs of a ladder, to the bunghole from which his guide, with the aid of 'a pretty instrument of some foote and a halfe long', drew up wine with which to 'exhilarate' his visitor. On this point, Coryate offers his reader a stern warning:

> I advise thee ... if thou dost happen to ascend to the top ... to taste of the wine, that in any case though dost drink moderately, and not so much as the

sociable Germans will persuade thee unto. For if thou shouldst chance to over-swill thyself with wine, peradventure such a giddiness will benumb thy brain, that thou wilt scarce find the direct way down from the steep ladder without a very dangerous precipitation.[1]

Dangerous – or at least nervous – precipitations seemed to be a theme of Coryate's actual mountain ventures. Coryate did not call himself a climber, but he was proud of his lengthy and adventurous perambulations. He was nicknamed (or, more likely, for he excelled at self-promotion, nicknamed himself) 'the Odcombian leg-stretcher' in honour of his home village of Odcombe, Somerset. He was an eccentric, a wit and self-appointed court jester to Prince Henry, the youngest son of James I and VI.

At first glance, all seventeenth-century books seem serious with their densely typeset pages, heavy paper and dark leather bindings. Thomas Coryate's 1611 book, *Coryate's crudities hastily gobled up in five moneths trauells*, quickly dispels any such preconceptions. The title played on a double meaning: the 'crudities', alluding both to things that were crude or unformed, but also to *crudités*, sticks of raw vegetables to be 'gobbled up'. The ornate frontispiece – effectively the illustrated 'front cover' of early modern books like this one, albeit contained inside the leather-bound boards – depicts key moments in his absurd adventures, such as sitting in a gondola and being bombarded with eggs by a woman leaning out of a window above. The opening pages of the book include verses written by his friends, supposedly in recommendation of the volume, but mostly making jokes at the author's expense.

Coryate's journeys took him across Europe, to the eastern Mediterranean, through Persia and finally to his death in India. His *Crudities* relate his European travels which passed through the mountainous region of Savoy (then a Duchy and a country in its own right and now incorporated into the modern-day borders of France). He met with a mountain mishap, which he retold with much relish, in climbing 'the Mountaine Aiguebelette, which is the first Alpe' en route to the town of Chambéry.

Now, in modern usage there is no peak by this name, although Lac d'Aiguebelette is noted as one of the largest natural lakes in France and is recommended today as an attractive tourist spot due to its blue-green waters and hot water springs. This lake, in Coryate's account, is termed 'an exceeding great standing poole' beside a 'poore village', where he and his companions paused to refresh themselves before launching their ascent. Moreover, despite his frequent references to climbing to the top of a mountain, it seems likely that Coryate in fact crossed through one of the three passes along the Chaîne de l'Épine, the ridge dividing Lac d'Aiguebelette and Chambéry. The highest of these, and the closest to Chambéry, is the Col de l'Epine, at 987m, or 3,238ft.

Coryate was not a confident horse-rider, and he lent his horse to one of his companions, for he deemed it 'more dangerous to ride than to go afoot … but then this accident happened to me': he and his group of travellers were accosted by 'certain poor fellows who get their living especially by carrying men in chairs from the top of the hill to the foot thereof towards Chambéry'. These men 'made a bargain with some of my company, to carry them down in chairs, when they got to the top of the mountain'. Coryate – the Odcombian leg-stretcher, you must remember – disdained such measures, but was also conscious that these palanquin-bearers knew the way up and down the pass. The bearers likewise knew that Coryate did *not* know the way, and, 'they being desirous to get some money of me, led me such an extreme pace towards the top, that how much soever I laboured to keep them company, I could not possibly perform it.'[2]

Poor Tom: faced with a path of 'innumerable turnings and windings thereof', beset 'with an infinite abundance of trees' on all sides, and forced to follow in the hasty footsteps of local guides whose main goal was to exhaust him into paying them to carry him. At last, 'finding that faintness in myself that I was not able to follow them any longer, though I would even break my heart with striving', Coryate gave in, and paid them to carry him the final half mile to the top. He did not enjoy the experience one bit:

> This was the manner of their carrying of me: They did put two slender poles through certain wooden rings, which were at the four corners of the chair, and so carried me on their shoulders sitting in the chair, one before, and another behind: but such was the miserable pains that the poor slaves willingly undertook: for the gain of that cardecu [coin], that I would not have done the like for five hundred. The ways were exceeding difficult in regard of the steepness and hardness thereof, for they were all rocky … and so uneven that a man could hardly find any sure footing on them.[3]

As he descended from his chair, Coryate, like any good seventeenth-century courtier, quoted Latin to himself, thinking 'with *Æneas* in *Virgil*: *Forsan et haec olim meminisse iuvabit*'. This passage (*Aeneid*, 1.203) is a tricky one to translate. It comes from a speech given by Aeneas, the future founder of Rome, to his Trojan men. They were stranded on an island after losing thirteen of twenty ships to a storm at sea, having fled their city following the admittance of the Trojan horse and the pillaging and defeat of the city by the Greeks concealed within. (The *Aeneid* is, in modern terms, the 'sequel' to the *Iliad* of Homer, told from the side of the Trojans rather than the Greeks.) Modern translations

1. Thomas Coryate being carried to the top of an Alpine pass on a chair. Frontispiece of *Coryate's crudities* (1611). (CC-BY, reproduction kindly provided by the National Library of Scotland)

generally give some variation upon 'A joy it will be one day, perhaps, to remember even this', and the 1550 translation by Thomas Phaemer – which Coryate may well have read – gives the suggestion that, 'To think on this may pleasure be another day.'

Talk of joy or pleasure sounds like overstating the case in the original context of the *Aeneid*, given the trauma suffered by the Trojans. However, in the case of an eventful mountain ascent and when written by a man known for having his tongue in his cheek, I suspect that Coryate's Latin reference can be read as saying, 'one day I'll look back on this and laugh.' Indeed, ever ready to invite the laughter of others at his own expense, Coryate included a depiction of his ignominious ascent in the illustrated frontispiece to his *Crudities* (fig. 1). Even modern-day climbers less prone to self-parody may recognise the truth in Virgil's sentiment: the most unforgiving slog up and down through miserable weather suddenly seems more enjoyable in retrospect once dry, warm and fed back at one's tent or hut. There is even a modern term for this phenomenon: 'Type 2' or 'second degree' fun.

Once at the 'top' of the 'mountain' – or more likely the saddle of the pass – Coryate enjoyed some first-degree fun too. He was satisfied to discover that he could now 'justly and truly say, that which I could never before, that I was above some of the clouds', for though the mountain he was upon was dwarfed by its taller neighbours, he could nevertheless observe clouds gathering on the sides of the mountain beneath his feet.[4]

From Chambéry, Coryate travelled onwards, with mountains looming over him 'on every side like two walls'. Coryate admired the mountain springs rippling down the mountain sides, the wild trees of every variety, and the sight of a 'wondrous high mountain ... at the top whereof there is an exceeding high rock' a mile beyond Chambéry (probably the Nivolet, 1,547m). But the mountains were not wonders without danger: Coryate feared rockfalls, and again dismounted his horse when traversing precarious mountainside paths with long drops to one side, noting ravines 'four or five times as deep in some places as Paul's tower in London is high' and anxiously imagining himself at the bottom of them.[5]

Coryate's next mountain ascent was that of the Mont Cenis pass (2,081m). Crossing this pass, Coryate's attention was all for the view, and for the strange story attached to it. 'I observed,' he commented, 'an exceeding high mountain ... much higher than any that I saw before, called Roch Melow: it is said to be the highest mountain of all the Alps, saving one of those that part Italy and Germany. Some told me it was 14 miles high.'[6]

The mountain to which he referred was Rocciamelone. At 3,538m, it is far from the highest mountain in the Alps by modern reckoning. It is possible – given that the borders of the Holy Roman Empire, ruled by the 'German-Roman Emperor', contained Switzerland at the time – that the one higher mountain to which Coryate alluded was in fact Mont Blanc. Whatever its absolute altitude, Rocciamelone was a sight to see, 'covered with a very microcosm of clouds', all but 'a little piece of the top' concealed, and this appearing like 'three or four little turrets or steeples in the air' – a veritable castle in the clouds.[7]

Coryate dedicated several pages to relating a story that his local guide told about Rocciamelone. Legend or, as Coryate put it, 'a pretty history' told that 'a notorious robber' whose conscience suddenly struck him 'for his licentious and ungodly life' decided to carry two paintings, one of Christ and one of the Virgin Mary, to the top of the 'highest mountain of all the Alps', where he would spend the remainder of his life making atonement through fasting and prayer. He thus 'went up to a certain mountain that in his opinion was the highest of all the Alpine hills', and settled down for the rest of his days. Unfortunately, it seemed this robber-turned-mountain-dweller had climbed the wrong mountain, for two more paintings suddenly appeared to him, causing him to realise (Coryate was not quite sure how) that 'he had not chosen that mountain which was the highest of all'. He then wandered until he came to Rocciamelone, which he ascended and never came down. The story entertained Coryate, but he ultimately deemed it no more than a 'tale or figment'.[8]

Having tarried (in the pages of his book at least) for quite some time at the top of the Mont Cenis pass, Coryate then endured the experience shared by both A.A. Milne's 'Tigger' and many a modern mountain-hiker: that the coming down is always harder than the getting up. 'The descent of this mountain I found more wearisome and tedious than the ascent,' he complained, for rather than riding he was forced to walk on foot, along 'uneasy' ways 'all stony and full of windings and intricate turnings … I think there were at least two hundred before I came to the foot.' Despite the challenging route, Coryate was not alone on the path. He reported meeting many people ascending in the opposite direction, along with laden mules and even 'a great company of dun kine', or brown cattle.[9]

Throughout his travels, Coryate continued to toil up and down mountains, over ways rocky and uneven, and admired the sight of them. On reaching Heidelberg, he exclaimed that, 'the situation thereof is very delectable and pleasant. For it standeth *in convalli inter fauces montium*' – in the jaws of the mountains.[10] He noted that these hills were covered by vineyards which presumably produced the very wine which filled the barrel upon which we first met him.

So, whether ascending a mountain pass in a precarious palanquin, or wobbling atop a mountainously large wine barrel, Coryate's tongue was regularly in his cheek, his thoughts ever to relishing the experience in the moment and, when that was not possible, to marking the 'second order fun' which he was having. Did he climb and appreciate mountains in the same way a mountaineer of today would? Of course not. But like many of his fellow early modern travellers, Coryate did not have to be a mountaineer in order to enjoy the mountains.

Chapter One

Mountain Ventures and Adventures

Why visit a mountain? It is a question that today at least has a fairly obvious answer: to climb to the top of it. Or, perhaps, depending upon your sporting proclivities, to ski down it. If you posed that question to my early modern friends, however, you would find a much wider range of answers. As we just learned from Thomas Coryate, the reason was sometimes 'to get to the other side'. If you asked Conrad Gessner (a Swiss botanist) he might tell you enthusiastically that mountain slopes offered rich opportunities to gather and even discover rare plants. Gessner, who is probably one of the more well-known early modern mountain enthusiasts, would also add that one should visit the mountains 'for the sake of suitable bodily exercise and the delight of the spirit'. If you asked Jean de Thévenot, a Frenchman known chiefly for his travels, you would hear about mountains as being the objects of pilgrimage, which you might ascend halfway to get close enough to the place where Jesus was tempted by the Devil. The poet Francesco Petrarca would tell you with a shake of his head that you could ascend a mountain, but all you would discover at the top is how much you had forgotten God.

To a modern reader, particularly a mountaineer, these may all seem like poor excuses compared to the intention of going to a mountain in order to climb it 'for its own sake'. The reader may be even more unimpressed to learn that most of the above activities – crossing a pass, collecting botanical samples, praying at sites of pilgrimage – regularly left out the proper summit of the mountain entirely. And this is exactly what has happened in the past fifty years or so of history-writing about mountains in Europe: scholars have searched for accounts that reflect our modern motivations for visiting the mountains, descriptions of enjoyment that match modern sensations, and have come back empty-handed ... precisely because those motivations and sensations are modern ones. But just because people in the fifteenth, sixteenth and seventeenth centuries did not set out with flags in their backpacks, ready to claim a summit for country, club or self, does not mean that they did not venture into the mountains and did not have adventures – both hair-raising and hilarious – upon them. So let us leave our modern yardstick at home and go in search of a few premodern mountain stories.

A TOURIST TO THE MOUNTAINS

Jean de Thévenot (1633–1667) was born in Paris, visited countries across Europe and ventured further afield to Egypt, Palestine, Persia and India over the course of twelve action-packed years. His travels began in 1655 and ended with his death in Persia (after being accidentally shot with a pistol) in 1667 aged just 34. The first volume of his *Relation d'un voyage fait au Levant* was published during his lifetime in 1665, with a further two volumes published posthumously from his surviving travel journals. An English translation, *The Travels of Monsieur de Thévenot into the Levant*, was produced two decades after his death.

What caught my attention was Thévenot's account of his time spent in the Holy Land in 1658 and his visits to and ascents of mountains which held religious significance. The paths which he followed were well worn: pilgrimage to the Holy Land dates back to at least the fourth century AD. In his journal-style account Thévenot tells us a great deal about the practicalities of religious mountain 'tourism' during the seventeenth century. He also reveals a strange combination of modern-seeming pride in his physical achievements alongside a distinctly early modern lack of self-consciousness regarding his ultimate athletic 'failure' on the last mountain he attempted.

Thévenot's first ascent was of what he called 'the Mountain of St. Catherine', or Gabal Katrîne. This is the highest mountain in Egypt at 2,692m, and said to have been the temporary resting place of the martyred Saint Catherine of Alexandria (who was threatened with torture using a breaking wheel, which shattered when it touched her and is thus also called a 'Catherine wheel'). Like many of the travellers in this chapter, Thévenot wrote in the first-person plural ('we'), despite never naming or demonstrating any particular attachment to the fellow-members of his travelling caravan. Thus, he and his nameless cohort departed from their resting place, the Monastery of the Forty Martyrs, situated at the foot of the mountain at Deir al Arbain, at one o'clock in the afternoon, 'taking with us a little Arab boy, who carried a small leather bucket full of water, that we might drink when we were dry'. Their climb took three hours, by ways that were 'full of sharp cutting stones, and many steep and slippery places to be climbed up, that hinder people from going fast'.[1]

I often think of that 'little Arab boy', carrying a heavy water bucket up those sheer slopes whilst the adult travellers went unencumbered. Keep your eyes peeled, and you will find such figures in the background of many a mountain venture, whether taken in the sixteenth century or the twentieth. Sometimes they will be mentioned – guides, porters, Coryate's chair-bearers – but sometimes not at all, and the only hint of their presence is the sneaking suspicion that the visitor to the mountains would never have found their way up or through them without local knowledge.

In the history of modern mountaineering it is all too easy to point to these kinds of relationships, between the outsider with power or money and the mountain-dweller with the load on their back, and blame it on the recent history of Western colonialism or imperialism. That was certainly what I thought when I first looked into the Everest expeditions of the 1920s, which attacked the mountain (there: the language of imperialism) from the Tibetan side, and thus travelled through a region little-frequented by visitors, and over which the British Empire dearly wished to have control. I came to the uncomfortable realisation that whilst the heroic George Mallory and his companions did not actively mistreat the communities and porters whose resources and labour they benefited from, they did travel through the landscape with an assumption that all the food and assistance which they required would and should be forthcoming.

At the time, I thought the explanation for this unfortunate phenomenon was as simple and complex as 'the British Empire', which enfolded all who belonged to it in a comfortable rhetoric of the innate superiority of the British and their right to rule the world. The more I looked at earlier accounts, the more I realised that this relationship between the mountain visitor and the taken-for-granted local resounded throughout time. We will come back to the overlooked mountain-dweller later, and try to look at him or her a bit more closely on their own terms, rather than on those of the outsider.

The body of Saint Catherine was apparently transported to the summit of the mountain by angels following her execution at the hands of the Roman Emperor Maxentius. It remained there, according to Thévenot, for 300 years before being brought down the mountain by a band of monks to its final resting place in the Monastery of St Catherine, some miles to the east.

The first miraculous landmark which Thévenot and his companions came across was a pool of water, which had allegedly sprung from the side of the mountain to slake the thirst of the exhausted pall-bearers. It had frozen over and Thévenot and his companions irreverently (and unnecessarily, given the little boy carrying their water) bashed at the ice with their sticks before deeming it to be frozen solid. At the top of the mountain they reached a 'Dome', or cave, which had allegedly sheltered St Catherine's body before its retrieval by the monks. There they found, in Thévenot's words:

> a great piece of rock rising a little out of the ground, whereon (they say) the angels placed it, and it bears still the marks, as if a body had been laid on the back upon it ... The Greeks [the Orthodox monks] hold that this cave was made by a miracle, but there is some likelihood that it hath been done by the hands of Men[.]

His scepticism regarding the miraculous nature of the cave and its markings aside, Thévenot and his companions duly paid their devotions before descending 'with a great deal of trouble', returning to their base camp at the Monastery of the Forty Martyrs at 6 p.m., 'feeling tired enough' five hours after setting out.[2]

The next day, a Sunday, they set out at 7 a.m. in order to ascend what Thévenot called 'the Mountain of Moses', which was 'not so high, nor so hard to ascend' as Gabal Katrîne, although fairly well covered with snow. This was the hill upon which Moses apparently fasted for forty days and received the Ten Commandments from God.

※ ※

Above, I throw out modern-day place names as if they were obvious – Thévenot's Mountain of St Catherine is today's Gabal Katrîne, his Monastery of the Forty Martyrs is at Deir al Arbain. These correspondences, however, are not as straightforward as they might seem. Not only have place names changed since the seventeenth century, but so too has geography itself: the very assumptions of what different topographical terms mean have shifted, and names have not just been translated but moved, referring to different landscape features from century to century, rather like Thomas Coryat's Mount Aiguebelette, where now there is only a lake by that name.

For a long time, I was perplexed by Thévenot's 'Mountain of Moses'. The mountain he seemed to describe matches a peak now named Jabal Musa (Mount Moses) but also known as Mount Sinai. Thévenot did not use either of those names, and also referred to 'Mount Sinai' elsewhere. In a preceding chapter, he described 'our arrival at Mount Sinai'. Here, he referred to no specific peak, but simply to entering 'among high Mountains'.[3] I believe that this represents a shift in the usage of the apparently singular term 'mount'. Early modern texts often seem to use the term to refer to a mountain range or massif. Specific summits, by contrast, are 'Mountains': the Mountain of St Catherine, the Mountain of Moses. For Thévenot, 'Mount Sinai' was not a specific peak, but a region which encompassed several mountains, including the one now thought of as Mount Sinai.

His account of the mountain of Moses, and the area around it, caused me further bemusement. He described descending the Mountain of Moses by stone-cut steps leading to Saint Catherine's Monastery (which, to confuse matters further, is not at the foot of Gabal Katrîne), and commented that, 'One may judge the height of St. Catherine's Mount by this [mountain], which is certainly not so high by a third, and yet hath fourteen thousand steps up to it.'[4]

These steps still exist. Modern tourism websites are united in giving them a count of 3,750, terming them the 'Steps of Repentance', with the accompanying story that they were cut by a monk seeking remission for his sins. Thévenot did not mention this story, and somewhat overcounted the number of steps.

This can be read as straightforward exaggeration, a foible of many early modern travel writers, but his suggestion that Gabal Katrîne is three times the height of the Mountain of Moses takes more unpacking. Looking purely at elevation – height above sea level – Mount Sinai, or Jabal Musa, is 2,285m high, and Gabal Katrîne is 2,629m high – hardly three times again as tall. Another overestimation by Thévenot? It seems so, until one turns to the prominence of the two mountains, i.e. the peak's relative height to the landscape around it. The prominence of Gabal Katrîne is not far from its 'true' height, at 2,404m – those three hours of navigating 'sharp cutting stones' really was a lot of ascent for Thévenot, his fellow travellers and the little boy carrying their water. By contrast, Jabal Musa stands only 334m in prominence from the valleys around it – so, if anything, Thévenot was underestimating the difference in scale between the two ascents.

I think this tells us something very interesting about how Thévenot – and people in the early modern period in general – viewed height as compared to today. Now, having surveyed the Earth using modern trigonometrical methods, we define and order mountains in terms of their *elevation*: Mount Everest is the highest mountain in the world, K2 the second. However, when organised by topographical prominence K2 falls far down the list. Mont Blanc, Kilimanjaro and Aconcagua, to name but a few, are all 'higher' in terms of the metres of ascent involved.

Early modern 'natural philosophers' (or scientists – sort of) were interested in measuring the height of mountains and had some means of doing so even if they were not as precise as those of today.[5] However, for people 'on the ground', like Thévenot, what made a difference to them was not absolute altitude but the distance – both seen from afar and experienced underfoot – between the base of the mountain and its summit. Prominence, not elevation.

Thévenot offered me one final moment of confusion, and it is one that highlights the precarious relationship between cultural ideas and specific topographical locations. After walking round St Catherine's Monastery, he 'saw at some small distance, Mount Horeb, on which Moses fed his Flocks, when he saw the burning bush'.[6] This is obviously, as far as Thévenot is concerned, a *different* mountain from the Mountain of Moses upon which the eponymous leader of the Jews received the Ten Commandments. Most modern scholars take 'Mount Horeb' and 'Mount Sinai', as named in the Bible, to be the same peak, although they are not necessarily agreed in locating them at modern-day

Jabal Musa, even though the tourist literature today confidently claims both the Ten Commandments and the burning bush for the peak. Evidently Thévenot, or his guides, thought of them as two different locations. Precisely which prominence or rocky outcrop Thévenot was looking at and whether we would today define it as a 'mount' or 'mountain' is impossible to tell from the text.

⊰ ⊱

Thévenot came to the massif of 'Mount Sinai' on the second of February, climbed Gabal Katrîne, the highest mountain in Egypt, on the third and trekked up and around Jabal Musa on the fourth. At the end of this, he declared himself 'very weary after so much mounting and descending', and grateful that they had enjoyed the mountains without wind for, 'whether hot or cold, it would have kill'd us.'[7] Thévenot clearly felt no need to present himself as a heroic or hardy climber.

On 15 April, not far from the city of Jerusalem (the spiritual summit of all pilgrimages), Thévenot and his companions came to the foot of what he called 'the Mount of the Forty Days Fast', or 'the Mount of the Quarantine'. Today it is known as the Mount of Temptation, and though it rises a mere 366m in elevation it dominates the skyline of modern-day Jericho with its rocky, pyramidical mass. This was the mountain upon which Jesus was said to have spent forty days and nights fasting and where he was repeatedly tempted by the Devil.

Thévenot's account of this peak is contradictory. He opened by criticising previous accounts of the climb up Mount Quarantine, insisting that, 'It is not so hard to go up, as some have been pleased to say.' He quickly followed this, however, by asserting that in some places it was 'very dangerous, for one must climb with hands and feet to the rock, that is smooth like marble'. At the time of his ascent it had recently rained, so the rocks were slippery, 'but we assisted one another', and so reached a grotto within which Jesus was said to have sheltered during his time on the mountain. It was here, only partway up the mountain, that Thévenot stopped, left behind by his companions who sought to visit another site of holy significance:

> Some of our company went up to the top of all the hill, to the place where the devil carried our lord, and tempting him, showed him all the

kingdoms of the Earth, saying, *All these will I give thee, if thou wilt fall down and worship me.*

Why did Thévenot go no further? He explained it quite baldly and without embarrassment; he was 'spent and weary', and the way was dangerous, with paths 'not two foot broad', with 'a great precipice to the side of it'.[8] He quite simply did not feel like it.

In this sense, Thévenot offers a striking contrast with mountaineering accounts of today. Can you imagine reading an account of a modern climber turning their backs on a summit so phlegmatically and unapologetically? It seems that the very top of a mountain did not hold the same siren call in the seventeenth century as it does today, even when it offered the temptation of standing in the footsteps of Christ. More than that, Thévenot – though elsewhere he emphasised the height or difficulty of the mountains he did ascend – seems to have felt no compulsion to demonstrate his personal physical prowess or to blush at simply feeling too tired to go on.

One final feature of Thévenot's mountain adventures that intrigued me immensely from the very start is the fact that they reveal what can only really be called a thriving and somewhat corrupt tourist trade centred upon the hills and peaks of the Holy Land. Visiting the River Jordan (where Jesus was baptised) and Mount Quarantine, Thévenot joined a group of travellers allegedly 4,000 strong, who together paid for a force of 500 soldiers to guard them from dangers on the road. The convoy was commanded by a *müsellem*, an Ottoman cavalry officer.[9] The group consisted of both 'Greek' (Orthodox) and 'Latin' (Catholic) Christians.

At the foot of Mount Quarantine, the *müsellem* set up a tent and began counting the travellers who wished to visit the mountain. Thévenot and his immediate travelling companions were the guests of the Convent of St. Saviour, in Jerusalem, and their costs to visit the pilgrimage sites around Jerusalem were being covered by the monks. When Thévenot and his five companions passed through the tent, the *müsellem* counted eight pilgrims to be charged to the convent's account. This was not, however, the worst extortion practised upon travellers to Mount Quarantine. Certain groups, including Catholic travellers like Thévenot, had special permission to ascend the peak. Others, such as the Greek Orthodox pilgrims, did not. The *müsellem* allowed the latter group to pay the initial fee and ascend the mountain anyway, but on their return bound them in cords and demanded further recompense for having turned a blind eye before they would be released.

Throughout his mountain travels, Thévenot was also cynically aware of the intervention of human hands to make locations of note just that little bit more notable. In addition to his doubts regarding the impression of Saint Catherine's body at the summit of the mountain, he also rolled his eyes at the mark of a camel's foot on the rock below the monastery of Saint Catherine on Mount Sinai. This footprint was said to have belonged to the camel of the Muslim prophet Muhammad. Thévenot noted that travelling Muslims kissed the footprint with 'great devotion', theorising that the Orthodox monks 'have made it to captivate their friendship, to the end they may reverence those places' – the pecuniary value of visitors to pilgrimage sites clearly making little distinction in terms of creed.[10]

Elsewhere on the Mount Sinai massif Thévenot and his companions visited the site where the Israelites, doubting that Moses would ever come down from his mountain ascent, briefly abandoned their faith and raised up an idol to worship. Visitors like Thévenot could peer into the rock and see 'a great head of a calf cut to the life', within which 'as the Greeks say', the golden calf was cast from the earrings and bangles of the Israelites. Once again, Thévenot was dubious; much more likely, he thought, that the local monks had cut the calf into the stone themselves.[11]

Thévenot's suspicions did not stop him visiting and making note of these physical markers of long-vanished events, and it is fascinating to consider these traces of a very real 'tourist infrastructure' surrounding the mountains which Thévenot climbed. Guides and water-bearers for hire. Stories told enthusiastically by the resident monks who, one can presume, at the very least solicited a small donation for their hospitality. Footprints carved into rocks to (depending on your viewpoint) either appeal to the credulous or provide a focus for sombre devotions. A whole squadron of guards to ferry pilgrims on their circuit around the environs of Jerusalem, and a well-oiled system of profit-making and extortion. The mountains of the Holy Land were well and truly the places to be in the seventeenth century.

THE RIDICULOUS AND THE SUBLIME

Imagine the scene: you are a 30-year-old Scotsman, on pilgrimage to Jerusalem, and you go for a swim in the River Jordan. You see an appealing tree on the side of the river and decide you would like to cut down a branch to make into a hunting stick to give to your king. You climb the tree stark naked, during which time your travelling companions – a group of Franciscan friars and other pilgrims escorted by Ottoman soldiers – leave you behind, before being attacked by a group of bandits. You view the fight from your treetop – agonising over whether you should feel grateful to have escaped the battle or obliged to go join in – and eventually sprint, leaving your clothes behind, to join the group once the bandits have been dispersed. One of the soldiers, exasperated by your vanishing act, tries to hit you with his pike and one of the friars not only intervenes but also throws you his gown so that you might conceal 'the secrets of nature' from your fellow pilgrims.[12]

Of all the mountain travellers I have encountered in my research, William Lithgow (c.1583–c.1645) is probably my favourite – although I would not invite him to a dinner party that I wanted to end in anything other than chaos. He was the kind of character for whom the term 'colourful' is made: he apparently had his ears cut off in early adulthood by the brothers of a young woman with whom one can only assume he was discovered in a compromising position. He was born in Lowland Scotland, the son of a merchant, and allegedly tortured in Spain as a suspected spy. As one biographer puts it, Lithgow 'seems to have attracted adventures to himself like iron filings to a magnet'.[13] Like Thomas Coryate, Lithgow prided himself on making most of his journeys by foot and claimed to have covered more than 36,000 miles under his own steam.

Lithgow's many footsteps inevitably led him up and down hills and mountains. Like Thévenot, he visited Mount Quarantine, which 'by the computation of my painful experience', he estimated to be 'above 6 miles' in height. Unlike Thévenot, Lithgow made it to the very top, proudly pointing out that he was but one of only eight members of a much larger company that dared the ascent. Up this winding, tortuous way (to a mighty altitude of 366m, remember!), Lithgow endured 'diverse dangers, and narrow passages' to reach the spots where 'they say' Jesus fasted and was tempted by the Devil. On the descent, another adventure occurred, although Lithgow leaves us in the dark regarding the precise details:

> In our return again, we had a most fearful descending, for one Friar Laurenzo would have fallen five hundred fathoms over the rock, and broke his neck, had it not been for me, who rashly and unadvisedly endangered my own life for his safety ... To recite all the circumstances of his deliverance ... I purposely omit to avoid tediousness.[14]

In his own vague words, Lithgow was thus the hero of the hour.

❧ ☙

The tall tales of William Lithgow inspire an important question. This is the question of truth and how much it matters in the study of the past. In a genre like travel writing, so often written to entertain and sometimes – as one suspects in the case of Lithgow – to aggrandise the author – how much can we rely on the events described as having actually happened? And if they didn't happen, does that mean the text should just be discarded, ignored?

In the academic field of history, this question of truth can most straightforwardly be summed up with reference to two ideas: positivism and postmodernism. Historical positivism is the stance that it is very much possible to retrieve the truth about the past. In a positivist viewpoint, history is almost a science: the data is out there, and the role of the historian is to reveal it and let it speak for itself. Positivism is a very comforting approach to history, I think; in a world in which the present day feels ever-shifting and amorphous, it is a reassurance to possess a firm and unchanging past upon which to stand.

Opposite to positivism stands postmodernism and its cousin poststructuralism. These theories, when applied to the study of history, introduce an enormous amount of doubt into our ability to ever retrieve the 'truth' about the past. They emphasise that what we have left of the past – mostly written sources – are unreliable: they are not the past itself but rather records produced by fallible authors, to be read hundreds of years later by equally fallible readers. Different types of records focus on certain things and leave other things out. For example, even dry governmental tax records are 'biased' in that they record a very specific set of data. Authors writing for different purposes and different audiences might consciously or unconsciously emphasise, conceal, or even lie about certain aspects of the 'facts' recorded. Even an ostensibly 'private' diary could be written with posthumous readers in mind or be subject to the deceptions that we practise upon ourselves about our own behaviour and motivations.

At the most extreme end, a postmodernist approach to history effectively says that all we can tell about a written source is that it represents what someone possibly (but possibly not) thought was the truth when they wrote it. Modern-day readers of past texts are equally unreliable. We will focus on the details that interest us most, gloss over those which do not, and interpret ambiguous passages based on our own perspectives and presumptions.

I vividly remember my first encounter, as an undergraduate, with positivism and postmodernism, and a heated exchange that I had with a friend and fellow history student. I found the sense of uncertainty about the past not only convincing but exciting; he thought it was all nonsense. Of course we could know 'the truth' about the past. I have since touched upon the same question with my own students and have come across much the same division of attitudes. Of course, insists one undergraduate, we can unpick what a writer in the past really meant; we can be scientists of the past, wholly empirical and impartial. Then, on the other hand, there is the student who laughs at that attitude, who says, how arrogant that is! Isn't it exciting to see ourselves as one more in a long line of people attempting to understand the past, and adding our own impressions to it?

At an academic level, the study of history today generally falls somewhere between the two poles of positivism and postmodernism. Historians must believe that we are uncovering some truth of some sort, otherwise there would be no point at all to our hours spent poring over archives or (increasingly in this age of digitisation) ruining our eyes staring at computer screens. At the same time, historians today are increasingly interested in inherently subjective topics and the unstable sources which can reveal them: what did people feel in the past, what did they think about this thing or that thing?

So for me personally, it doesn't really matter whether William Lithgow actually once saved one Friar Lorenzo from a sticky end on a mountainside. What matters is that he thought it was the sort of thing worth including in his self-congratulatory book, whilst at the same time it matters that Jean de Thévenot did not blush to admit that he didn't fancy climbing to the summit of the self-same mountain. It gives us a glimpse, however partial, of the varying contours of seventeenth-century mountain heroism, so very different from that of today.

As well as having a penchant for tall stories, William Lithgow was also an aspiring (this is a key adjective) poet. With his literary aspirations in mind he paid particular attention to Mount Parnassus (2,457m), which was said to have been the home of the Greek Muses.

The importance of the ancient Mediterranean world to the wider intellectual life of the seventeenth century cannot be underestimated. The inheritance of classical Greece and Rome shaped the way mountains were viewed, too. I would even go so far as to say that in terms of cultural prominence, Parnassus was one of the 'tallest' mountains in early modern European thought. What do I mean by cultural prominence? Well, if you were to stop someone on the street today and ask them to name the first mountains that come into their head, they would probably offer a list including Everest, K2, Mont Blanc, the Eiger, maybe Ben Nevis, if they were Scottish, or Snowdon if they were Welsh. These are all tall mountains, mostly famed today for the perceived physical challenge of ascending them. I think if you had stopped a moderately well-read person on the street in the early 1600s and asked the same question, they would have said: Parnassus, Etna, Mount Athos, and the Peak of Tenerife (more on this surprising inclusion shortly). Parnassus was a peak that would have leapt readily to the tongue of any educated early modern European prompted to 'think of a mountain'.

Lithgow viewed Parnassus from the side of his boat as it sailed along the Thessalonian shore during a journey towards Constantinople (today known as Istanbul). Parnassus is a double-peaked mountain with a saddle in between. One of the two tops, Lithgow observed, in a somewhat self-pitying mood, was 'dry, and sandy, signifying that poets are always poor'. The other was 'barren, and rocky, resembling the ingratitude of wretched, and niggardly patrons'. The col between the two tops, however, was 'pleasant and profitable, denoting the fruitful, and delightful soil' to be worked by poets, 'the Muses' ploughmen'.[15]

You can judge for yourself how fruitful that soil proved for what Lithgow himself termed his 'poor poetical vein'. In 1615, on his way home from visiting Africa, he travelled via Sicily and another classically famed mountain: Mount Etna. Ancient myth associates various stories with the ever-active Etna: one story tells that Zeus threw the mountain at the giant Typhon when he rebelled against the gods, and that its grumblings and eruptions mark the struggles of its prisoner to escape. Another identifies the peak as the forge of Vulcan, the divine blacksmith. Lithgow reported that he ascended the volcano through 'tedious toil, and curious climbing', despite the danger presented by the 'terrible flames, and cracking smoke'.[16] He claimed to have been inspired to compose, on the very spot, a sonnet to the fiery mountain:

> High stands thy top, but higher looks mine eye,
> High soars thy smoke, but higher my desire:
> High are thy rounds, steep, circled, as I see,
> But higher far this Breast, while I aspire:
> High mounts the fury, of thy burning fire,
> But higher far mine aims transcend above:
> High bends thy force, through midst of Vulcans ire,
> But higher flies my spirit, with wings of love:
> High press thy flames, the crystal air to move,
> But higher far the scope of my engine:
> High lies the snow, on thy proud tops, I prove,
> But higher up ascends my brave design.
> Thine height cannot surpass this cloudy frame,
> But my poor Soul, the highest Heavens doth claim.
> Meanwhile with pain, I climb to view thy tops,
> Thine height makes fall from me, ten thousand drops.[17]

In the introduction, I wrote about the old canard that 'people didn't like mountains back then'. Part of the change which is assumed to have come about with the modern era is the development of the idea of the sublime in nature. It has generally been thought that this only really formed part of the European experience of mountains from the eighteenth century onwards, when philosophers like Edmund Burke and Immanuel Kant articulated a specific aesthetic response to things that were great or awe-inspiring rather than merely 'beautiful'. In brief, a flower or a woman could be beautiful, but the sublime could only be experienced through the sight of something immense, like a mountain or the night's sky. A critical aspect of the sublime was also the ability of the rational human being experiencing it to be awed but not overwhelmed, with the suggestion that the human mind was in fact 'greater' even than the highest mountain peak.

Looking past Lithgow's somewhat forced rhyming, all of these layers are evident in his poetic description of Etna. He looked at the volcano and was impressed by its scale, the strength of its flames, but always he went higher. Ultimately, he implied, the mountain could never reach beyond its earthly limits, however great they might be; whereas he, a Christian faithful enough to risk his life and certainly his clothes for a dip in the River Jordan, would ultimately ascend to a heavenly reward. At the same time, his concluding lines admitted the power the mountain had over him whilst in the mortal realm. The ascent was one of pain, and he sweated out 'ten thousand drops'. This poem captures precisely the contradiction that can be found in many books about modern mountaineering: an uplifting sense of awe and sublimity, rubbing alongside the lowering, physical reality of climbing a mountain; the aches and pains and drops of sweat.

DOING HISTORY: MY BOOK

The physical reality of 'doing history' involves rather less sweat than climbing a mountain, and aches and pains in different places. These days, the primary complaints of the historian are back pain and eye strain from poor posture and a surfeit of 'blue light' as a result of hours spent hunched in front of a computer. It is both a blessing and a curse that, these days, it is entirely possible to study a period like the seventeenth century without ever having to handle an actual physical book or document; digitisation projects such as Google Books mean that a vast array of historical texts are available online. There is no need to travel to distant libraries to find a physical copy of a rare volume, or to peer through a magnifying glass at densely printed lines of text. Now all it takes is a few words typed into a search engine and you have a PDF, which you can set at whatever level of zoom you prefer. Even when historians do visit archives, these days it is generally with a camera or more likely a smartphone to hand, with the goal of photographing material as efficiently as possible, so that in a day's visit they can collect images of pages that would take weeks to read through. The vast majority of the 300- or 400-year-old books that I have consulted over the past decade of research have been in the form not of physical objects but digital files.

There is one exception, which is at my elbow as I type these words. It is a chunky volume, some 2 inches thick, 11 inches tall, 7½ inches wide. Even closed, you can tell it has been through the wars: the leather binding on the front and back covers is chipped and worn, the edges of the pages dark with centuries of accumulated dust and grime. The spine, which was replaced at some point, proclaims the date of publication as 1662. At almost four centuries old, then, the book can be forgiven for looking a little battered. Opening it, the browning title page reads: *The Voyages and Travels of the Ambassadors from the Duke of Holstein, to the Great Duke of Persia. Begun in the year M. DC. XXXIII. and finish'd in M. DC. XXXIX. Containing a compleat History of Muscovy, Tartary, Persia, And other adjacent Countries. With several Publick Transactions reaching neer the Present Times; IN SEVEN BOOKS. Illustrated with diverse accurate Mapps and Figures.* The author is given as Adam Olearius; the English translator one John Davies of Kidwelly, a Welsh town in Carmarthenshire.

Turning the pages, one finds further signs of heavy use: messy scribbles adorn the back of the title page and one of the largest of the 'diverse accurate Mapps' has been painstakingly rebacked with a sheet of cotton to prevent it falling to pieces. On the back of various plates, one finds the signature of a previous owner. John Batcheller, with spiral flourishes to the 'J', is proudly swashed across the back of a plate depicting the face of the Michael Federowitz, the Duke of Muscovy. On the back of a map, in smaller letters, we find 'John Batcheller His Book Steyning Sussex 1770', and the same assertion of ownership on another map a sheaf of pages later. Finally, on the back of the last map included in the volume, John's name has found a partner, with quick, satisfied pen-strokes proclaiming 'John & Mary Batcheller 1773'. The first time I saw this, I decided that the childish scribbles on the back of the title page surely belonged to the toddler progeny of the Batchellers, perhaps trying out his father's pen in around 1776, to be greeted with the same exclamation of resigned exasperation – why did I leave *that* there? – of the modern-day parent who finds their child scrawling across the wallpaper with biro.

The year before I submitted my PhD, 'John Batcheller His Book' became 'Dawn Hollis Her Book'. I had won a university prize for an essay about, and catalogue of, my collection of modern mountaineering books, most of which had been picked up in second-hand bookshops around the country for a few pounds apiece. The prize was a few hundred pounds, to be spent expanding my collection. I immediately wondered whether I might finally be able to fulfil an ambition of owning one of the titles which appeared prominently in my PhD research. Most copies of Olearius sell in the thousands; this battered example, however, had left the bookseller unimpressed by its prospects on the open market, and they had priced it just within my reach. So it was that when it came time to double-check my references to Olearius throughout my thesis, I did so not squinting at a PDF but turning real pages. Every time I need to be reminded that I am a historian, not a mere cruncher of digital data, I pull this volume down from its pride-of-place upon my shelves, stroke its battered leather, and smile at John and Mary and their imagined book-vandalising offspring. It reminds me that the material I write about was not merely recorded in one moment, simply for me to pluck it from the internet in another moment hundreds of years later, but in between took physical form and was read and treasured and scribbled upon by countless individuals before it came to me.

⁂

And what is contained within John Batcheller's book – my book? The travel account of one Adam Olearius (1599–1671), sometime mathematician, geographer, librarian and, for the years 1633 to 1639, secretary to a pair of ambassadors travelling through Russia and Persia. The ambassadors were in the service of the Duchy of Holstein, one of the states of the Holy Roman Empire (and which now forms part of modern-day Germany) seeking treaties with Tsar Michael of Russia and Shah Safi of Iran, the latter of whom was known for being addicted to opium and heavy drinking, but also for abhorring tobacco so much that he sentenced anyone caught smoking it in public to death. Olearius's account was first published in German in 1647 and first translated into English in 1662. Identifying the exact summits that Jean de Thévenot climbed is a walk in the park compared to taking the place names in Olearius – transliterated almost 400 years ago from Cyrillic, by a German – and accurately pinpointing his location on a modern, anglicised map.

The main goal of an embassy is to get to places of power – urban centres, cities. However, in an age before flight, to get to these places Olearius and his employers had to cross vast swathes of sparsely habited landscapes, by both river and land. During their travels, Olearius was continually alert to the sights and scenes around him. Travelling down the Volga from Kazan to Astrakhan – a journey of almost 1,000 miles – he peered at the cliffs and mountains rising to either side of their vessel. In the vicinity of the modern-day cities of

Tolyatti (Тольятти) and Samara (Самара) the Volga describes an inconvenient loop, almost doubling back on itself entirely. It narrows and any easy route for boats is complicated by small islands and sandbanks. The peninsula created by the inner circuit of this loop is craggy – no more than 200 or 300m high, but in some cases rising almost sheer from the water's edge. Olearius made note of several 'mountains' – and that is the word he uses – around this area. The first one they passed served as a salt-mine, with huts for processing right on the very mountainside. The second mountain, on the other hand, provided the perfect lookout for robbers lying in wait for travellers on the road below. The vulnerability of visitors was often on Olearius's mind: several times during their journey down the Volga he and his companions noted suspicious riders on the roads and paths along the waterside.

One peak in particular caught Olearius's eye. He termed it 'Diwisagora', or 'the Maids' Mountain', for, 'the Muscovites say it derives its name from certain Maids that had sometime been kept there by a She-Dwarf.' In a modern transliteration, this maid's mountain would more properly be written as Devichya Gora (Девичья Гора), but there is no peak today known by that name along the Volga, although there is one near St Petersburg. There is a folk tale from the Urals, recorded in the early twentieth century, of the 'Mistress of the Copper Mountain', which may bear some distant relationship to Olearius's mountain maid. The folk tale tells of a beautiful, green-eyed woman, sometimes called the 'Malachite Maid', who acted as a sort of patron saint to miners, which is certainly intriguing in light of the mining activity which Olearius observed. Like the sand banks of the Volga, the associations between stories and specific landscapes shift, often irrecoverably, through time and space. Whatever its legends, 'Diwisagora' was an appealing mountain:

> very high and steepy [sic] ... pleasant to the eye by reason of the diversity of the colours, some being red, some blue, some yellow, etc. and representing, at a great distance, the ruins of some great and magnificent structure.[18]

Olearius travelled down the Volga and admired the many-coloured (even if now frustratingly anonymous) mountain of the maid in August of 1636. December of that year found him on the shores of the Caspian Sea in Persia in the vicinity of what is today the border between Georgia and Azerbaijan. On Christmas morning, he and his fellow travellers awoke in a village recorded in the *Travels* as 'Schamachie', now either vanished or linguistically transformed beyond all recognition. They gathered together in the 'great stable' where the camels were kept and 'did our devotions'. Religious duties performed, some of the party 'had the curiosity to go and take a view of the mountain' behind the village. This peak, Olearius stated, was of an 'extraordinary height', visible from the sea from a great distance, and of such steep rock that, 'it looks like a finger stretched out above the other adjacent mountains.' Olearius gave the name of this mountain as *Barmach*. After hours of squinting at modern-day maps (and, perhaps a touch misled by his description of its 'extraordinary height', looking to the great chain of 4,000m peaks running parallel to the shore of the Caspian Sea) I discovered that the shift in spelling from the 1600s to today was not as significant as I had expected, for the peak he climbed would today be termed Besh Barmag, or Beş Barmaq, which is 382m high and with a name which translates as 'Five Finger'.

Despite its modest height, it is as imposing a figure as its name suggests, and Olearius reported ruefully that since he and his companions missed the actual path to the summit, they instead 'ran great hazard of our lives in getting [up it] by dreadful precipices' – an experience surely familiar to any mountain hiker who, seeking an easy shortcut to the summit, finds themselves anxiously off-piste. On top of the mountain, the tall grass 'was all covered with a white frost as [if] with sugar candy'. The group rested on a rock, sang *Te Deum*, a Latin hymn of joyful praise to God, and 'renewed among ourselves the friendship, which we had before mutually promised each other by most unfeigned protestations'. The language here is flowery, but given that the preceding pages made multiple references to receiving gifts of wine and aqua vitae from various local hosts, I think I can translate: the diplomatic gentlemen opened a bottle of something strong and slapped one another on the back as they drank it, dizzy with the height and good feeling of Christmas Day. They then gathered some figs from trees growing between the rocks of Beş Barmaq, and 'got down again with less trouble and danger, by the ordinary path'.[19] An exciting ascent, refreshment at the summit, and an easy descent – all in all a pleasant Christmas leg-stretcher.

August to December 1636 thus took Olearius from gazing up at mountains from the (relative) comfort and security of a river-boat to ascending a rocky pinnacle by pathless ways for festive entertainment. Over a year later, Olearius found himself on a more challenging mountain journey. They had spent several months in Isfahan (Olearius spelt it 'Ispahan'), then the capital city of Persia

– long enough for one of the ambassadors in Olearius's party to bring their diplomatic endeavours into disrepute by marrying local women (plural). They made their journey back towards the Caspian Sea in January 1638, stopping at Qazvin ('Caswin') before crossing the Western Elburz Range, a distance of over 40 miles even at the narrowest possible spot. The route they took, Olearius reported sombrely, was 'the most dangerous and most dreadful way of any, I think, in the world'. In some parts, the rocky mountainside was so steep that past travellers had used tools to widen the path sufficiently to allow a camel or a horse to pass through. The noise of the river and the length of the drop to one side shook Olearius and even gave their experienced guides some pause:

> On the left hand, the rock reached up into the clouds, so as that the top of it could not be seen; and on the right, there was a dreadful abyss, wherein the river made its passage, with a noise, which no less stunned the ear, than the precipices dazzled the eye, and made the head turn. Not one of us, nor indeed of the Persians themselves, dared ride it up, but were forced to lead their horses by the bridle, and that at a distance loosely, lest the beast, falling, might drag his master after him.

In the end, no horses were harmed in the making of this early modern travel account, and the whole party safely reached the summit of the pass, where they paid their due at a toll-house. As they descended from the top of the pass a change came over both the landscape and the mood of the travellers:

> But this mountain, which was so steepy [sic], tedious, and dreadful on the one side, had so pleasant and delightful a descent on the other, that it was no hard matter for us to forget the fright and trouble we had been in, in coming it up. It was all over clad with a resplendent verdure, and so planted with citron [lemon] trees, orange trees, olive trees, nay, cypress trees and box, that there is not any garden in Europe could more delight the eye, nor more surprise and divert the smell. The ground was ... covered with citrons and oranges, in so much that some of our people who had never seen such abundance of them, made it their sport to fling them at one another's heads.[20]

Reading the story of this fruit fight, on the heels of thoughts of aqua vitae at the summit of Beş Barmaq, I strongly suspect Olearius and his diplomats would have made fairly pleasant hiking companions. I also know I am not the first person to read, in my particular copy, of the dreadful ascent and delightful descent of the Elburz mountains. For at the bottom of this very page, there it is again, in smaller letters this time: the signature of John Batcheller.

MODERN OR NOT?

I have a question to put to you. What is the highest mountain in the world? You may find this a strange question for me to ask, so obvious is the answer. The highest mountain in the world is of course Mount Everest, 8,848m high. But is a mountain the highest mountain in the world if nobody knows it exists? If you take the line that a tree falling in the woods makes no sound unless someone is there to hear it, then you might agree with the apparently fabulous claim I am about to make: Mount Everest was not always the highest mountain in the world. At least, not so far as my European friends knew.

If I were able to time travel and ask the same question to a person on the street in seventeenth-century London, they would also think my question absurd. The highest mountain in the world was obviously Mount Teide, also known in the early modern period as the Peak of Tenerife. At 3,715m this volcano is the highest point of the largest island of the Canaries and the highest point in Spain as a whole. I wrote earlier about the distinction between prominence and elevation in terms of Thévenot's definition of the relative heights of peaks in the Sinai peninsula. Mount Teide suffers from no such qualification: an island peak, its base begins at sea level. Every metre of its 'real' height, its elevation, is visible as prominence to the traveller approaching by sea, perhaps one reason for its seventeenth-century status as the highest mountain in the world.

There are numerous accounts of individuals ascending the Peak of Tenerife. Perhaps the most charming is that of Marmaduke Rawdon (1610–1669), who lived on Tenerife as factor, or business manager, for his uncle from 1631 to 1655. One summer, he decided to fulfil his desire to ascend the highest mountain in the world, and so in the company of fifteen other gentlemen from England, the Netherlands and Germany he set out with servants and 'mules laden with wine and provisions'.[21] Snow stood on the mountain top all year round, and when the party bivouacked for the night the wind blowing from the summit was so cold that Rawdon and another Englishman stayed up all night building fires to stay warm. Their guide urged them to begin their march at 4 a.m., and they all walked on foot, in purpose-made shoes used by local goatkeepers to grip the rock. Rawdon climbed well. His biographer, writing either during Marmaduke's lifetime or shortly after his death, reported proudly that some of the party managed only 'a quarter part of the way, some half the way, and could get no further; but Mr Rawdon got up very well and was the second person upon the Peak, there being only one German gentleman before him'.[22] Those that made it tarried at the summit for an hour or so, enjoying a panoramic view – apparently able to see as far as 150 miles all around – and the sight of 'the clouds lying like fleeces of wool under them'.[23]

Before they left the highest point in the world, however, Rawdon had a trick up his sleeve:

> Whilst they stayed here Mr Rawdon called their guide, being a lusty proper fellow, the tallest of all the company, gave him a piece of money, and told him that he would have him to take him upon his shoulders, and that after that he should take up none else, which he promised him to do; so, when he was set upon his shoulders, looking about him he said to the company, 'I am now the highest man in the world, and the nearest heaven of any man living.'[24]

You can just imagine Rawdon's wide grin and the eye-rolling, probably, of his companions at his proclamation of his own cleverness. This scene is immensely interesting because it is an example of something you do not see a great deal of in early modern accounts: the desire to be 'the highest', to reach the very 'top' of a mountain. It is something we take for granted in modern mountain ascents: no one wants to report that they had to turn around before reaching the summit of Everest (although many are forced so to do). Yet, as we found in Thévenot, seventeenth-century climbers saw no shame in stopping before the summit, and travellers such as Thomas Coryate and Olearius happily defined the saddle of a pass – really the lowest convenient point at which you can cross a mountain – as 'the top'. In some ways, Rawdon feels far more familiar to us today with his desire to be able to proclaim 'I'm on top of the world!'

And here is the problem: it is all too tempting, even easy, to focus on what is familiar in history, what is recognisable. It is very human: we like to find others who are like us. I also think – and I will elaborate on this in the final chapter – that we are, unconsciously, very proud of being 'modern'. Forays into the history of mountain engagement have generally been conducted under the assumption that the love of mountains only truly came to be with the start of the modern era. However, investigations into the 'prehistory' of modern mountaineering have tended to pay attention to a select few accounts which seem at first glance to resemble the way we experience mountains today, whilst discarding those which do not. This has resulted in the illusion that mountains were only mentioned, or climbed, very rarely in the premodern period. It also runs the risk of those few accounts being viewed entirely through the lens of 'how modern' they were, taking them out of their own rich and specific historical context.

Rawdon has received little attention of this type, perhaps because his apparently 'modern' conceit is buried within a floridly written and distinctly early modern biography. There are, however, a pair of early modern characters worth discussing in terms of their modern mountain fame: Francesco Petrarca, and Conrad Gessner.

Long ago, I asked myself whether it was possible to write a book about early modern mountains without writing about Francesco Petrarca, the Renaissance Italian poet and letter-writer more commonly known in the Anglophone world as Petrarch (1304–1374). Petrarch ascended Mont Ventoux (1,912m) in 1336, and for this feat has received several accolades. The nineteenth-century Swiss historian Jacob Burckhardt termed him the 'first modern man', whilst elsewhere his ascent of Mont Ventoux is described as the 'birthday of Alpinism'. But what did his famous climb look like?

Recorded in a letter to his former confessor – the Catholic priest to whom he declared his sins and received absolution – his adventure started with a book of ancient history.[25] Titus Livy had written of the ascent of 'Mount Haemus' (the premodern term for what is now known as the entire Balkan mountain range) by the elderly King Philip V of Macedon in the second century BC. Petrarch was inspired by the image which Livy gave of the ruler taking in the view of two different seas, the Adriatic and the Euxine. Looking at the landscape around him and settling upon the highest mountain in his region,

Petrarch decided that 'a young person might attempt a feat that an old king was not criticized for undertaking'.[26]

Petrarch's first concern was for his choice of travelling companion. Mentally, he flicked through a catalogue of his friends. One talked too much, one talked too little, another was too slow and another too fast. The flaws that were acceptable at ground level would, Petrarch privately thought, rapidly become intolerable on an arduous mountain journey. He finally found his ideal partner closest to home, in the form of his younger brother Gherardo. (Readers with siblings may, as I did, raise an eyebrow at this conclusion.)

At the foot of the mountain, they met with an elderly shepherd who, when he heard what they were planning, launched into a rant:

> Fifty years previously, he said, the same youthful ardour had driven him to mount the final summit, and he had gained nothing from it but regret and fatigue, as well as torn clothes and scratches from the rocks and thorny bushes.[27]

The more he harangued the two brothers, the more determined they became to attempt the feat. Finally, the shepherd gave up, pointed out the path for them, and took custody of their excess baggage so that they might ascend unencumbered by all but the necessities.

At this point, it became apparent that Petrarch had not chosen the best-matched climbing partner after all, for his brother forged a path directly up the mountainside. Francesco, the elder and apparently weaker, waved him goodbye and said he would try to find a less direct but easier route. In the end, he found himself 'wandering in the hollows' for hours, and when he finally rejoined his brother on the ridge of the mountain, Gherardo was refreshed from a long rest waiting for him, whilst Francesco was 'exhausted and weak'. As the youngest of three siblings, and thus often the slowest in childhood on a family hike, I will wager a bet that as soon as his brother reached his side Gherardo leapt up, declaring himself ready to continue, whilst Francesco gritted his teeth, too stubborn to insist on a rest for himself.

Such misadventures – Gherardo leaping ahead like a mountain goat, Francesco toiling up tortuously convoluted routes behind – apparently repeated themselves several times on their path to the summit, 'not without,' Petrarch commented glumly, 'making my brother laugh and me angry.' At one point, the older brother sat down, feeling momentarily defeated. Along with a rest, he gave himself a pep talk which compared his physical journey to his spiritual one: 'Don't forget that the same trials that you ... endured today, will also be met on the road to blessedness.' He exhorted himself with the words not of any Christian author but rather the Roman pagan, Ovid: 'To want counts little; to triumph, you must ardently desire.' For all that he might wander in the low foothills of the spirit, Petrarch told himself, he would eventually 'climb to the heights of blessedness'. Having turned his time on the slopes of Mont Ventoux into an extended metaphor for his journey towards heaven, Petrarch stood up, newly strengthened for the final ascent of the physical mountain before him.[28]

At the summit at last, he was struck dumb. 'I looked around,' he wrote, 'and the clouds floated around my feet.' He could see Italy, the country which he loved, and 'Bristling and snowy, the Alps themselves where formerly the fierce enemy of the name of Rome [i.e. Hannibal] opened a passage by breaking the rocks with vinegar'. He could see the bay of Marseilles, waves crashing on the coast, and the Rhône flowing 'beneath our eyes'. The sight filled him with 'earthly joy', but then he thought to pull a book which he had carried with him from his pocket: the *Confessions* of Augustine. (Before becoming a Catholic priest and eventually a saint, Augustine had stolen fruit for the fun of it in his childhood, fathered an illegitimate son in his youth, and famously prayed to be granted 'chastity and continence, but not yet'.) Francesco opened the book at random, and his eyes fell on lines that seemed to be a message direct from the erring saint to the mountain wanderer: 'Men go to admire the mountains and the immense waves of the ocean and the vast courses of rivers and the circuit of the sea and the revolutions of the stars, and they abandon themselves.' Petrarch was 'stunned', and he did not look out at the view again or speak until he and his brother reached the foot of the mountain.[29]

The remainder of the letter sees Petrarch bemoaning the vanity of human nature. How foolish human beings were (or rather, as he put it, 'men'), who would 'go looking outside for what they can find within themselves', and how foolish he was for wasting his sweat and passions merely to carry his physical body 'a little closer to heaven', when his real focus should have on been dragging his soul out of the mud of human passions and temptations.

Petrarch's account has been the subject of substantial scholarly discussion over the decades, including those which have queried whether he made the ascent at all or whether it was in reality written as an allegory for his spiritual journey (rather than an experience which he actually had and compared to it).[30]

Nevertheless, the myth of Petrarch as the first modern man, as the father of Alpinism, has persisted; as one history of climbing put it, he was 'a real mountaineer, that is to say someone who made the climb for pleasure alone'.[31] This vision of Petrarch overlooks entirely his spiritual summit epiphany, his horrified rejection of his original enthusiasm to climb the mountain, and his deliberate turning away from the view around him. Yet, to reduce his account to an analysis of which parts 'were modern' and which 'were not' prevents us from seeing them as part of the same whole. His pocket-book Augustine, his self-flagellation, helped form the same premodern mountain experience which also encompassed striving for the summit and delighting in the view.

Conrad Gessner (1516–1565), who wrote and published about his experiences of mountain climbing in his homeland of Switzerland, is a similarly tempting example of a 'modern' feeling for mountains appearing centuries ahead of time. One early member of the Alpine Club, in a particularly poetic vein, termed him 'the morning star of mountaineering'.[32] Gessner was – like many notable early modern writers – a miscellany of things in his lifetime: a naturalist, a bibliographer, a philologist, an illustrator. He also had ambition. His *Bibliotheca* (1545) sought to catalogue all the writers who had ever lived, his *Historia animalium* (1551–1558) offered a 4-volume, 1,500-page encyclopaedia of animal life, and for his *Historia Plantarum* (unpublished until long after his death) he collected almost 1,500 botanical drawings.

Claimed as the father of modern-day botany, zoology and bibliography, Gessner also features in many histories of modern mountaineering for two – by his standards – relatively short texts. The most oft-cited is his 'On the admiration of mountains', published in 1541, though it is sometimes paired with his description, published in 1555, of 'Mount Fractus', also called (and known today as) Mount Pilate or Mount Pilatus.[33]

A feature of early modern books that may seem curious to the modern eye is the tendency of authors to include prefatory letters, often addressed to high-status individuals (such as a monarch or a well-known scholar) in the apparent hope that the implied association would help promote the work. The modern equivalent is probably books with forewords written *by* other such individuals, but in the case of early modern books these opening letters sometimes diverged wildly from the actual subject matter of the book they prefaced. If anything, they were sometimes used as excuses by authors to get their thoughts on loosely related topics into print.

This very much seems to be the case with Gessner's 'On the admiration of mountains', a letter addressed to a friend named Jakob Vogel. It formed the preface of a treatise titled *Libellus de Lacte, et Operibus Lactariis* – On Milk and Milk Products. Gessner rather tenuously justified the connection because his friend Vogel lived in a mountainous area which also produced cheese. The letter opens with a phrase for which Gessner has become, in the annals of mountaineering, relatively famous. 'I have resolved, my friend Vogel, for as long as God grants me life, to climb some few mountains – at least one – every year when plants are in fresh growth, for the sake of their study as well as for the health-giving exercise and spiritual enjoyments [that] mountains afford.'[34]

In his account of climbing the peak now known as Mount Pilatus, Gessner expanded further upon the joys of mountain climbing. There was no greater pleasure, he claimed, than to stop for a drink of cold springwater when parched from the exertion of ascent. Indeed, he insisted that every sense received its own reward for venturing into mountain country: a climber's sense of touch was variously cooled by the wind or warmed by the sun, and their sight 'delighted' by the diverse colours, shapes and overwhelming magnitude of the mountain landscape. One's hearing was 'entertained by pleasant talk of friends, jests, and turns of wit', and by the singing of birds in the woods. The mountains represented an escape from the buzz of urban life: 'There is nothing to annoy; there are no disturbances or city fracases, no disputes of men.' The sense of smell was rewarded by the 'pleasant fragrances of herbs, flowers, and mountain greenery', and by the absence of the noisome and unhealthy vapours believed by many early modern writers to envelop towns and cities. Returning to taste, Gessner found mountain fruits to be more pleasing than their counterparts from the plains, and yoghurt and cheeses made from mountain herds to be 'far and away more outstanding'. More than that, any type of food was more pleasing after the hard labour of clambering up rocks than after a day of sitting around at home. 'And so let us conclude,' announced Gessner, 'that from mountain ramblings undertaken with friends the highest of pleasures and the most satisfying of all sensations are to be enjoyed.'[35]

Gessner followed this proclamation by addressing himself to an imagined critic of mountain sport. This caricature protested that 'walking itself' was unpleasant, and could perceive only the dangers of 'high places and precipitous drops' in the hills. (Such complaints are by no means confined to early modernity: my older brother, when dragged up the Brecon Beacons as a teenager, repeatedly complained 'but what's the point?') To such a person, Gessner would say – echoing the same ancient words of Virgil that Thomas Coryate would allude to sixty years later – that, 'it will be a joy afterward to remember these labors and risks, and to turn these things over in the mind and recount them to friends.' But what about, whined the anti-mountaineer, the fact that 'one has to go without bed, pad, feather blankets, and pillows!' Gessner had little time for such scruples: 'Hay will suffice for all of this! Hay will be your pillow, your mattress piled all beneath you, and as a blanket spread over you.'[36]

In all this, Gessner seems a modern mountaineer displaced in time. He could not let a year pass without climbing a mountain; he delighted in the physical exertion of doing so and shook off as trivial, even actually enjoyable, the potential challenges of walking, eating and sleeping on the side of a mountain. Yet there is much that Gessner wrote about mountains that go unmentioned in texts that claim him as a 'true ancestor' of modern mountaineers.[37] His letter 'On the Admiration of Mountains', after opening with his declaration of making an annual ascent, really focused on the physical make-up of mountains, of fathoming out how they came to be within the terms of sixteenth-century knowledge. Centuries before any understanding of tectonic movement or glacial erosion, Gessner concluded that all mountains had been formed through eruptions of 'fire', such as could still be seen at work on active volcanoes such as Mount Etna. He also listed the many mythological figures and creatures – the god Pan, the nymphs, the Muses – located in the mountains in Greek mythology, concluding that they reflected a longstanding sense of the wonders of the mountain landscape. The mountains he named in his letter were Olympus, Etna, Vesuvius, Parnassus and Helicon, rather than any Alpine peaks.[38]

In his account of ascending Mount Pilatus, he further demonstrated his distinctively early modern approach to the mountain landscape. In arguing for the enjoyment of the climbing experience he referred to Epicurus, a classical philosopher who posited that what was pleasurable was also morally good.[39] In his response to the imaginary critic's anxieties about the dangers of mountain climbing, he commented that those with a fear of heights or a poor sense of balance could easily avoid steep cliffs, perhaps stopping halfway: 'It is something to have gone so far, even if one cannot go farther.'[40] Just as with Thévenot, getting to the very top of a mountain was not the be all and end all.

In his description of the summit area of Mount Pilatus, he barely commented on the view, showing more interest in the fact that former visitors had left graffiti on the highest pinnacle of rock. By contrast, he dedicated several paragraphs to the myths surrounding a mountain tarn encountered on the descent. This pool of water was said to be where Pontius Pilate – the Roman governor of Judea who, in the Bible, sentenced Jesus to death by crucifixion – drowned himself. Locals, Gessner wrote, believed that if anyone disturbed the pool, for example by throwing a stone into it, then 'the whole region would be threatened by storms and floods', such was the power of the malevolent spirit of the long-dead Roman. Gessner was sceptical. For one thing he doubted that the governor of Judea ever made it to Switzerland, and for another, he thought it impious to suggest that anyone but God should have the power to reward or punish such behaviour as lake-side rock-throwing.[41]

To me, both Gessner and Petrarch are prime examples of why it is not at all helpful to look at past mountain climbers in terms of the ways in which they were or were not 'like us'. Whilst some aspects of both men's mountain experiences do seem familiar, seeing them as somehow modern, or even as the precursors to Alpinism, means ignoring all of the fascinating and important ways in which they were *not* modern. It is also anachronistic: neither Petrarch nor Gessner experienced their ascents as 'sometimes modern' and 'sometimes not'; they were always and entirely their own experiences, informed by the contexts within which they lived – not the labels later historians would put on them.

And so, I ask you to take all of the stories from this chapter on their own terms. What matters is not that they depict a relationship with mountains that was either 'different' or 'the same' as ours. Of course, they were different – they were from a different time. And where they were the same, we miss the richness of that different time if we pay most attention to points of similarity. How did early modern travellers, early modern *climbers*, experience mountains – whether they climbed for pilgrimage, for a view, or just because a mountain was the thing in between them and their next destination? They got tired and gave up, they kept going and took risks, they admired the view and enjoyed the physical sensations of touch and taste and smell. They also viewed them as places of ancient myth and as places to connect with God – the latter sometimes prompting spiritual turmoil and sometimes a Christmas Day summit toast.

However, as the next chapter will show, poets and diplomats were not the only people to frequent the mountains of early modernity. In fact, these privileged visitors often relied upon the aid of the 'real mountaineers'.

Vantage Point: A Chapel in the Mountains

I am on my honeymoon in the Tarentaise Alps, on the south-eastern border of France. Italy is a mountain pass away, the tip of Mont Blanc just visible if you look north on a good day. It is late summer, and we drink the local wine by night and take easy hikes by day. One day, we take a walk to Le Monal, a mountain hamlet inaccessible by vehicle and whose eighteenth- and nineteenth-century chalets have been untouched by time, and seem small in contrast to the buildings of the nearby ski resort of Sainte-Foy, designed to host hordes of modern-day sporting tourists rather than pre-industrial families. On the path up to the hamlet, a spring of fresh mountain water disgorges itself into a basin with a small sign, '*eau potable*'. To drink straight from this spring is a pleasure almost equal to the pétillant wine of the region. In the middle of Le Monal is a *fromagerie*, a pastoral vision with goats in the back garden. Sparkling clear waters run through the hamlet – a large stream, or a small river. A large, periwinkle-blue butterfly lands on my hand as we sit and eat bread and cheese by the river. Flowers abound. I am struck by the appeal of the mountains in summer, and how much the warm, perfumed air contrasts with the classic, modern image of the Alps as snow-bound winter heights. In this spot it is easy to imagine the inhabitants of two or three centuries ago enjoying their own summertime, stepping out of their homes to milk their goats, washing clothes from the flower-strewn banks of the stream, and quenching their thirst at the mountain spring.

We walk to the far end of the hamlet, and begin to climb again, up a winding path which ultimately disgorges us into the higher valley of Le Clou, which looks like nothing so much as a bowl scooped out of the heart of a cluster of bristling mountain summits, the remnants of their ancient glaciers seeming almost close enough to touch. Here, the past is still on hand, but not so idyllic or nostalgically vivid as in Le Monal. A few ruined huts can be seen, memorials to an early modern alpage – a settlement used in the summer only, when herdsmen would bring their goats and cattle to enjoy the high-altitude grasses, rich and untouched over the winter.

Turning south, we climb up out of this mountain basin. Ten minutes from the hut stands an arresting sight (fig. 2). The footprint of a now-vanished and tiny building, perhaps 6ft by 10ft, with a strange sculpture opposite the door: an undulating stela, with a cross and a seashell carved into it. This is, our guidebook informs us, *l'ancienne Chappelle de Saint-Jacques*, and the sculpture marks the site of the altar, the seashell the symbol of Saint James to whom it was dedicated. Not many metres beyond the structure, the grassy mountainside falls away, and our eyes are drawn irresistibly to the other side of the valley, where the skyline is dominated by Mont Pourri (3,779m) and the massive glacier tumbling down its north-west side. On this warm summer day, it gleams bright and blue. My later research finds that the chapel stands at 2,200m above sea level, and was certainly standing in 1633 when a pastoral visit reported that it lacked both cross and ritual candlesticks and gave instructions for these to be purchased. It was still in use in 1778, when the parish priest climbed to the site to celebrate mass on the feast day of Saint James.

Fortunately for me, my new husband is also a historian, and though his interest in mountains is secondhand he waits patiently whilst I stand in the middle of the ruined chapel and stare all around me, struck by an entirely different way of thinking about the history of mountains. 'Ordinary' people are all too often left out of history, and to be honest had been left out of *my* history, so dependent on written records up to that point. Here there were only a few stones and a modern sculpture, but I could imagine the summer inhabitants of Le Clou, leaving their herds in the high-altitude basin below, making their way up to this col, to this chapel, so deliberately placed apart from their spaces of work and habitation. They could have built their chapel closer, amidst their chalets, as the chapel down at Le Monal had been built, but then they would not have seen this view. And what need had a chapel of fine candlesticks or even a cross to remind them of the glory of God when their last sight before entering and their first sight on leaving was of the high peaks, the frozen glaciers, and the Alpine meadow-flowers at their feet?

In the cool air, above the summer heat of the pastoral idyll down below, I think to myself with a conviction that has not been shaken since: the cheese-makers, the herdspeople, the mountain-dwellers of Le Monal and Le Clou did not for a moment fear or detest this landscape.

2. The chapel of St James above Le Clou in the Tarentaise Alps, France. (Author's photograph)

Chapter Two

The Real Mountaineers

The term 'mountaineer' is a tricky one. What does it mean? Today, most people would probably agree that the word refers to someone who climbs mountains. It conjures a particular image: a person on a steep slope, perhaps with an ice-axe in hand, probably dressed in brightly coloured Goretex outerwear, with a large rucksack on their back. A sportsperson. This usage, however, is a relatively recent one, dating from only around the mid-nineteenth century, when mountain climbers applied the term to themselves.[1] If you come across a text from before that point which speaks of 'mountaineers', it will most likely be referring to the people who lived in the mountains. The title is still claimed by some mountain-dwellers today. I was delighted, when attending a symposium in North Carolina, to hear men and women living in the Appalachians term themselves 'mountaineers' in the sense that would have been instantly understood by my seventeenth-century friends.

So my question in this chapter is this: what was it like to be a mountaineer in the fifteenth, sixteenth and seventeenth centuries? What did they do on the mountains, how did they make a living from them? What did they know about mountains, in the sense of being familiar with their dangers and having techniques to navigate them safely? And what did they think and feel about mountains? Did they fear the mountains and shrink from them? Or did they, as my instincts told me at the chapel above Le Clou, even take pleasure from them?

These are not necessarily easy questions to answer, not least because the real mountaineers left no written records of their own lives. Instead, we find them depicted in paintings intended to hang on the walls of wealthy home-owners, and find their doings recounted in letters, books and pamphlets written by people who visited the mountains. Those people were, generally, far more privileged than the mountaineers whose landscapes they passed through, and their accounts were frequently coloured by their assumption – both conscious and unconscious – of their own superiority and distinctiveness from the 'locals' or 'peasants' upon whose guidance and labour they regularly relied. To fully unpack how much of an impact this had on the 'accurate' presentation of the real mountaineers in historical records, it is worth travelling forward in time from the main focus of this book, back to where I started my own journey as a historian through the mountains.

THE MOUNTAIN 'OTHER' THROUGH THE AGES

In 1922, the British Empire hoped to conquer not a country but a mountain: Everest, the highest point on Earth. This was the second year in a row that British climbers had made their way to the peak, travelling through Tibet, a country which had fascinated Europeans for decades with its solitude and resistance to outside visitors.[2] The first expedition, in 1921, had set out to explore the region surrounding Everest and to chart possible routes up its hitherto-unmapped slopes, and to make a summit attempt 'if possible'. This time, with maps in hand, the explicit aim of the expedition was to reach the summit. Three attempts were made between 21 May and 7 June. By the third attempt, both climbers and 'porters' – Tibetan and Sikkimese hillmen employed to carry much of the equipment required by the expedition up the mountain – were exhausted.

So it was that on 7 June George Mallory was climbing up the North Col, a pass at around 7,000m, with two other Europeans and fourteen porters. They were divided into four small teams, each roped together. The three 'mountaineers' went first, taking a route straight up the steepest slopes of the glacier at Mallory's instructions. Just before 2 p.m., he heard 'a noise not unlike an explosion of untamped gunpowder'. Although an experienced climber, Mallory had never before experienced an avalanche, but he knew exactly what the sound meant. The first group was only partially buried; the second hit by 30m of snow, and the third and fourth thrown off into a crevasse and buried. One porter had the fortune to have been tied to Mallory, and rapidly emerged unscathed; the second group also shook themselves free of the snow. Nine remained to be found, of whom two were rescued, six were dug out already dead, and one was never found. Two days later, Mallory wrote to his wife, still reeling from the event:

> It's difficult to get it all straight in my mind. The consequences of my mistake are so terrible; it seems almost impossible to believe that it has happened for ever and that I can do nothing to make good. There is no obligation I have so much wanted to honour as that of taking care of these men; they are children where mountain dangers are concerned and they do so much for us: and now through my fault seven of them have been killed.[3]

This is a tragic story, and at first glance Mallory's response seems unproblematic: he was clearly stricken with guilt and horror at the loss of life, and the schoolmaster-turned-mountaineer felt a keen sense of concern and responsibility for the porters under his authority. Looking more closely at Mallory's language, though, some details might give us pause. The porters were 'children', with apparently no knowledge of their own of the mountain landscape. The term 'coolies' was used throughout the eighteenth and nineteenth centuries to refer to unskilled Chinese or Indian labourers. Throughout the letters, diaries and official records of the early Everest expeditions of 1921, 1922 and 1924, the term was applied generically to all of the 'native' people engaged by the British to aid their assault against the mountain. It obscured the wide variety of ethnicities to which the porters belonged, and identified them first and foremost as subservient, unskilled labourers at the bottom of the mountaineering machine.

Mallory's letters are far from the only documentary record of the expeditions, and reading the diaries, letters and reports of the other climbers reveal the various stereotypes through which they viewed the people of Tibet. Sometimes these stereotypes contradicted one another. For example, the British expedition members often praised the physical endurance and hard work displayed by the porters who carried equipment and supplies up and down the mountain. In the next paragraph, however, they could be found decrying the 'Oriental inertia' or laziness of the porters who, after a long ascent and after pitching the tents of the 'Sahibs', proved slow to construct stone huts for their own overnight accommodation.⁴ They also described the Tibetans as a simple people. This had some positive connotations, in that it made them 'cheery and good-natured companions', but they were also, according to the British, problematically superstitious, demonstrating reluctance to ascend the glaciers because they believed them to be 'peopled by powerful evil devils'.⁵

The climbers also wrote critically of the filth and dirtiness of both the Tibetan towns and the people themselves, which stood in contrast to the stark white beauty and ice-clean air of the surrounding mountains. They found the streets to be 'knee-deep in filth' and recorded that the Tibetans hated baths. One team-member noted with irony in his published account of the 1922 expedition that, 'Only once did I see a Tibetan having a bath… I learned that the boy was the village idiot, and therefore not responsible for his actions.'⁶ Meanwhile, in 1921, George Mallory wrote home to his wife about a grand money-saving scheme he had dreamed up, which was to bring a youthful Tibetan porter home as a houseservant. Among the many points in young Nimja's favour, Mallory insisted that 'he is a clean animal'.⁷ That he felt the need to point this out at all suggests that the idea of the Tibetans as dirty was a deeply ingrained stereotype.

Academic historians love to set things within 'theoretical frameworks', in which we match phenomena with fancy terms and try to identify patterns across place and time. The framework within which the British mountaineers' response to the people of Tibet can most helpfully be set is that of Orientalism as developed by Edward Said in the late 1970s. For Said, 'Orientalism' encompassed the ways in which the Western world perceived and related to the Eastern world. In general, Orientalism characterises Western cultures as viewing themselves as superior to those of the East, which are stereotyped as primitive, barbaric, and in need of enlightenment and improvement (to be provided, quite naturally, by the positive and 'modern' influences of the West). Said's critique of Orientalism formed the foundation of what is now known as postcolonial studies. Postcolonial studies turns a critical – meaning analytical, not deliberately 'negative' – eye on colonialism and imperialism, and on the relationships between those exerting power and those upon whom power is exerted within these contexts. It pays particular attention to modern European history, such as that of the British Empire.

A concept which is central both to Orientalism and postcolonialism is that of the 'Other'. Unlike a lot of theoretical jargon, the 'Other' is a fairly straightforward term that does what it says on the tin. The East is the Other to the West. The poor are Other to the rich. The central thing to know about the Other is that it is a *construct*, often used by those with power to (mis)describe and (mis)represent those over whom they perceive themselves as having power. The Other is quite clearly at play in the ways members of the early Everest expeditions represented and responded to the people of Tibet. They travelled with a whole set of preconceptions and assumptions about what the Tibetans as the Other were like. These stereotypes informed their behaviour towards the Tibetans they met and how they described them when recording their experiences.

More recently, a scholar named Dibyesh Anand has identified some of the specific ways in which modern Europeans have 'othered' the people of Tibet. These include debasement (depicting the Other as degraded or bad), idealisation (emphasising, for example, the simplicity, purity, or happiness of the Other), and infantilisation (viewing the Other as childlike – and in need of parenting by the coloniser).[8] It is very easy to apply these ideas, as well as the concept of Orientalism more generally, to the early Everest expeditions. After all, they are theories developed to explain roughly this cultural context and historical period. However, as I delved further into the history of mountains before modernity, I was struck by how often the ways in which the twentieth-century climbers viewed the people of Tibet – as simple and quaint, lazy and dirty – could be found in earlier encounters between those who inhabited mountains and the elite outsiders who visited them.

Daniel Defoe – most famous today as the author of *Robinson Crusoe* – published a three-volume *Tour thro' the Whole Island of Great Britain* between 1724 and 1727. His account of the Peak District was largely cynical in tone: it had been celebrated in the preceding century as a place of 'Wonders', and Defoe largely dismissed the caves, springs and hilltop which made up the list of seven wonders which informed the itinerary of most visitors to the area. (As an aside, I was entertained to recall, on reading Defoe, that I had visited many of the early modern 'wonders' as a small child on a twentieth-century family holiday – regional tourism clearly not being an industry apt to change rapidly.) However, he did admit that some of the *people* he encountered in this hilly landscape were worthy of attention.

Climbing Mam Tor (517m) in search of a cave then termed 'Giant's Tomb' and now called Giant's Hole, Defoe's travelling party came across a family living in a cave of their own. The gentlemen questioned the woman of the 'house', astonished by the family's habitation. Over the course of the conversation, Defoe was impressed by many things: that the woman possessed fine manners (for she addressed her visitors as 'sir' and 'your worship') despite living 'in a den like a wild body', the fact that the children were 'plump and fat, ruddy and wholesome', and that the woman herself was 'well shaped, clean, and ... comely'. The home itself was well kept, though the family poor; Defoe questioned her closely on this point and discovered that she and her husband between them earned at most 8 pence in any given day.

Before leaving, the visitors banded together to hand the woman 'a little lump of money', which astonished her as more money than she had seen all together 'for many months'. The story of this family, Defoe hoped, would 'show the discontented part of the rich world how to value their own happiness, by looking below them, and seeing how others live, who yet are capable of being easy and content'.[9] This family who dwelt inside a mountain were held up by Defoe as the ideal of the happy, deserving poor.

Not long after leaving the cave with its pleasingly plump brood of children overseen by a surprisingly clean and comely woman, Defoe's crew passed through a valley filled with narrow mine-shafts, or 'grooves', and stopped to stare at them. As they did, they saw 'a hand, and then an arm, and quickly after a head, thrust up out of the very groove we were looking at'. The head belonged to a man described in quick succession as a 'poor wretch', a 'subterranean creature' and 'a most uncouth spectacle', and as appearing like an inhabitant of hell 'who was just ascended into the world of light'. His accent was so strong that their guide had to translate for the ears of the upper-class southerners.

Once again, the travellers took the opportunity to spread their munificence, asking the miner if they could each take away a small piece of the lead ore which he had brought out of the ground. For this, they gave 2 shillings, apparently a vastly excessive price, for the miner informed them that such a sum would normally take him three days of underground labour to earn. This was not their final encounter with the 'subterranean creature', for the group later arrived at a local alehouse and found him working hard, as Defoe put it, to 'melt' his earnings into beer. Defoe deemed it a stroke of luck on the miner's part that their paths had crossed once more, for they bought him a pint and then sent him home to his family before he could spend the rest of his money.[10]

In Defoe's two encounters with people who quite literally *in*habited the mountains of the Peak District, he demonstrated the trifecta of forms of Othering that I found so interesting in Dibyesh Anand's article about Tibet. He idealised the 'comely' woman, living in her mountain cave with her plump children. More than that, he held her up as an example to, and a striking contrast with, the wealthy, civilised part of the world – to which he himself belonged – and the failure on the part of some of its members to find contentment despite all of their advantages. The miner, he 'debased', depicting him as (at first glance) a denizen of hell itself. On closer acquaintance, however, Defoe found the man to be a mere child – infantilising him by presenting him as in need of the indulgent guidance of a gentleman such as himself.

If the hillspeople of Derbyshire were a little strange, but still worthy of charity, the Highlanders of Scotland were deserving of admiration mixed with wariness – at least for Defoe. Like many writers of the era, he identified a family resemblance, as it were, between the landscape and the people who inhabited it. Just as the Grampians were harsh, wild and rocky, so too were the people, 'a fierce fighting and furious kind of men'. Once the Highlanders had been very wild, but according to Defoe they were now 'much more civilised than they were in former times', leaving behind a mountainous 'Vigour and Spirit', which meant that they made excellent soldiers, whether in the British army or serving as mercenaries abroad.[11] The exception to the rule of the hardy but civilised Scot came, for Defoe, in the example of the inhabitants of the Western Highlands, who were 'some of the worst, most barbarous and ill governed of all the Highlands of Scotland'. These men (and, really, Defoe was thinking of men) were 'desperate in fight, cruel in victory, fierce even in conversation, apt to quarrel, mischievous, and even murderers in their passion'.[12]

Defoe, travelling at the start of the eighteenth century, allowed that *most* Highland Scots were relatively civilised, but many earlier, seventeenth-century, English writers did not feel so generously towards their northern neighbours. The worst offender in this regard was Thomas Kirke (1650–1706). Kirke was born in Yorkshire, and attended but failed to graduate from the University of Cambridge. In 1677, he undertook a three-month tour of Scotland, and in 1679 he published – under the pseudonym of 'an English Gentleman', *A Modern Account of Scotland*. The subtitle to this seventeen-page pamphlet termed it 'an exact description of the country, and a true character of the people and their manners'. It was blisteringly, uncompromisingly rude.

The Scots, according to Kirke, were 'proud, arrogant, vainglorious boasters, bloody, barbarous, and inhuman butchers'. By his report, they regularly indulged in incest, stole from each other, and – in a passage reaching impressive levels of hysteria – allegedly engaged in a practice whereby they removed tasty cuts of meat from their cattle whilst they still lived. Their food was foul, their bagpipes an appalling excuse for music. Highlanders even had the temerity to dare speak their own language and look blankly at a visitor expecting English – no matter how hard the civilised visitor took at them with a cudgel.[13]

The only possible compliment that Kirke could find to give the people he encountered whilst travelling the Highlands was that the nobility took great pride in offering hospitality to strangers, but even this disgusted Kirke. It was, he claimed, 'their way of showing you're welcome, by making you drunk', depicting the unwary visitor as a 'conquered victim' left 'grovelling' after having too much beer, wine and sherry forced upon them.[14] This particular example of Highland degradation intrigued me – I had noticed similar criticisms in the writings of the Everest mountaineers regarding the excessively liquor-fuelled hospitality of twentieth-century Tibet. In both cases, the criticism seemed to be the height of hypocrisy. Alcohol – ranging from whisky to champagne – was included in the expeditions' stores, and one does not imagine that Kirke, over 200 years earlier, made much of an effort to say 'no, thank you' to yet another glass.

All in all, Mallory, Defoe and Kirke offer just a few examples to illustrate an important point. In terms of the written record, we generally find the inhabitants of mountains being written about by people coming into the mountains from outside. Such accounts – whether written about 1920s Tibet or 1670s Scotland – tend not to show us what the people in mountains were really like. Instead, they reveal the stereotypes, prejudices and assumptions held by the 'incomer', who viewed those who inhabited the mountains as inherently different from themselves, wild products of a wild landscape, worthy perhaps of sympathy but rarely of true respect. We therefore must not take such reports at face value. It does not take much thought to conclude that the seventeenth-century Highlanders certainly did not torture their cattle, but we might also take with a grain of salt the comment of George Mallory that men who had grown up among the high hills of Sikkim and Tibet were 'children where mountain dangers were concerned'.

An Inhabited Wilderness

One of the questions I asked at the start of this chapter was 'what did people do on the mountains in the early modern period?' This question takes as its starting point the fact that the mountains of this time were full of activity, of people coming and going, travelling the high roads, and putting food on the tables of homes nestled amidst the peaks. This runs counter to several assumptions made about the mountain landscape during the modern era.

One such assumption is the idea, prevalent in landscape history, of mountains as a 'marginal landscape'. This concept conjures up a sense of insecurity and privation for humans dwelling in such landscapes, and indeed of an absence of habitation unless absolutely necessary: who would want to live on the margins? However, it alludes more specifically to the agricultural value of a particular environment: if the soil is poor for growing crops, then it is a 'marginal landscape'.

The identification of a landscape as marginal if it lacks agricultural potential has its roots in the eighteenth century and the age of 'improvement', which coincided with the Agricultural Revolution in Britain. During this time, elite landowners turned to their land with an eye to how they could make it as economically productive as possible, taking advantage of the scientific developments of the Enlightenment. This resulted in decisions that have shaped the landscapes of today: for example, Scottish landowners of the late eighteenth century observed that they would do far better from their low-yielding hillsides if they emptied them of small tenant farmers and filled them with sheep, one of the first phases of the infamous Highland Clearances.

The legacy of the Clearances is that the Scottish Highlands – like many mountain landscapes around the world – seem to be quite empty of people. This has contributed to another modern idea associated with mountains, namely that of 'wilderness'. A wilderness is generally thought of as a place beyond human interference and habitation – or at least, that is the ideal. In the late nineteenth and early twentieth centuries, the conservation movement in the United States – led by figures such as John Muir, who was born in Scotland but lived most of his life in America – set the trend towards the modern-day protection of wilderness areas. Landscapes deemed to be areas of natural beauty find themselves designated as National Parks, with attendant limitations on the amount and type of human development permitted in these areas. Many mountainous areas now fall under national or international protection as wilderness spaces, to simultaneously be preserved for the enjoyment of humans, but protected from the impact of them.

The example of the Highlands, and the Clearances, should be enough to give us pause about this. The idea of a 'wilderness' captures a sense of a place that is and has always been natural, somewhere that is untouched and should remain so. But many modern 'wildernesses' are as constructed as any urban jungle. The Scottish Highlands of today are wild and uninhabited because wealthy landowners a few hundred years ago made them so. The idea of a wilderness obscures the fact that many spaces we now view as 'natural' and untouched have a long history of human habitation.

This is by no means a novel perspective. It was first put forward by William Cronon, a prominent environmental historian, in an article published in 1996. 'The Trouble with Wilderness: Or, Getting Back to the Wrong Nature' argued that, 'far from being one place on Earth than stands apart from humanity', wilderness 'is quite profoundly a human creation'. Falsely celebrating and advocating for spaces empty of humans is problematic, Cronon concluded, because it leaves no space in environmental thinking for 'what an ethical, sustainable, *honorable* human place in nature might actually look like'.[15]

For my part, I do not believe that the landscapes we now define as marginal wildernesses – especially mountains – were seen or experienced as such during the early modern period. The mountains of early modernity were productive, inhabited, and even busy spaces.

First of all, mountains were spaces of movement, with people travelling up and over them to travel between valleys. Today, in an age of vehicular travel, mountains often present challenges to easy movement. The modern response to this has been brutally practical, with nineteenth- and twentieth-century railway and road builders blasting tunnels through the Alps to achieve shorter journey times and avoid the inconvenience of high-altitude winter conditions. But the mountains of early modern Europe were places not of tunnels but of passes. Josias Simler (1530–1576) noted that in sixteenth-century Switzerland the local mountaineers were actively involved in maintaining the passes: 'Every day, men of the neighbouring villages on each side of the slope explore the path towards the col … and repair the path.' Following heavy snowfall, the villagers would lead their cattle up towards the pass. The beasts would not only break the snow with 'their legs and chests', but would also drag after them a long pole, presumably attached in such a way as to be perpendicular from their direction of movement, to flatten out swathes of snow behind them.[16]

Mountains and hills were not just landforms to be passed through but also highways in their own right. The cattle mentioned above formed a significant portion of the worldly wealth of any early modern mountain-dwelling farmer. However, in order to transform a head of cattle from a mass of living flesh and muscle into silver and gold, or other more transportable goods, the herd must first be brought to market. In early modern Britain, cattle were raised across the 'marginal' landscapes of Scotland, Wales and northern England, but they earnt the best price if taken, fresh and breathing, to bustling urban centres such as London, and their ever-growing populations. And if you were an early modern drover, responsible for moving several hundred large animals across the length of Britain, and keeping them fed, what route did you take? The modern answer would probably be, the easy route – stick to the lush valleys, to flat, low-altitude roads and paths, and go up into the mountains only when they need to be crossed. The early modern answer was to take to the heights. The hillsides offered many advantages to the drover and his cattle. It meant they could avoid the towns and the irritation – both bovine and human – of forcing the square peg of a herd of animals into the round hole of a space of dense habitation, with its noise and crowds and paved roads which pained the cows' hooves. They also avoided expensive tolls, and instead took advantage of plentiful free wayside grazing.[17]

Cattle were not just driven across the relatively modest mountains of mainland Britain. Josias Simler, again, offers written evidence of large-scale movement of herds across the Alps, commenting that, 'each year, actually, a great number of cows and horses are herded over the transalpine regions from Switzerland and Germany into Italy.' These routes could even become dangerously crowded. According to Simler, drovers would arrange with each other in advance the times at which they would take large herds across narrower portions of path, to avoid the potential disaster of two groups of cattle or horses meeting and becoming sufficiently disturbed that they 'might run off the cliffs and die'. There were also, should plans go awry, 'rules which determine which of the two groups can stay on the path, and which must yield the right of way'.[18]

Cattle were not the only organic matter which needed to be transported across long distances and preferably away from areas of dense urban settlement. Hiking in remote areas of Britain, you may come across the occasional large, flat stone beside the path. You might even stop to eat your sandwiches upon it. It may be worn and moss-covered, and even seem to be a quite 'natural' part of the landscape, shaped that way by the action of long-vanished glaciers. It is quite possible you are right, but it is also possible that you have just eaten your lunch on a 'coffin stone', where people carrying the dead by long roads to their final resting place could pause, and rest from their burden themselves.[19]

Go into any town today and keep an eye on the road signs and you are likely to notice a depressing, but also darkly humorous combination: not far from the large 'H' denoting a nearby hospital, you will see the signs for the town's crematorium. This makes perfect sense: data from Public Health England in 2016 noted that almost half of all deaths occurred in hospital. These proximate pair of road signs represent the brief journey that many people will take after death. The majority of people in Britain also live in towns or cities: wherever you die, you will never be particularly far from a crematorium or burial ground.

This was not the case in early modern Britain. Not only did more people live – and die – in disparate rural settlements, the options for what to do with a body were also more limited. In sixteenth-century Britain, when you died you wanted to be buried in consecrated ground, generally by a church. Parishes could be large, particularly by the standard of foot travel, and some smaller churches lacked the licence necessary to conduct burials.[20] If you died of an illness in your farm cottage or an accident in the mine, you then represented a logistical problem not entirely dissimilar from the herds of cattle: your body needed to get from the A of your remote home to the B of your rightful burial place. And just as with the cattle, remote 'corpse roads' – which often meant taking the *high* ways, rather than the highways – offered multiple advantages.

The pall-bearers were more likely to be able to travel uninterrupted, and both the dead person's spirit, and the spirits of townspeople, were less likely to be disturbed by coming into contact with one another.

Living cattle and dead bodies, then, passed over the mountains, but entire communities also went up them. The historic 'alpage' or summer pasturage at Le Clou which I visited on my honeymoon is just one visible example of a practice that was widespread across medieval and early modern Europe, and that is something called vertical transhumance. Transhumance refers to the movement of livestock (and the people who tended them), from one place to another from one season to another. Vertical transhumance, by extension, specifies the movement of herds from a lower to a higher altitude and back again. It is a form of nomadism, but usually the two sites – the winter 'base' and the summer pasturage – are fixed and do not change from year to year. There are different words across European countries and regions to specifically refer to the high summer pasturage, which said together sound like a prayer: the *shieling* for Scotland, the *hafod* for Wales, the *seter* for Scandinavia, the *alp* or the *alpage* for German-speaking Switzerland. In Ireland, the upper pasture was termed a *buaile*, or *booley*, and thus the act of moving to it from the lower ground was called, wonderfully enough, 'booleying'.

Why, however, would you want to move from your main home at lower ground, up to your remote shieling or seter or alpage? There were several reasons, all of which demonstrate the ways in which communities quite ably adapted to the 'marginal' nature of the mountain environment. Once it was warm enough at altitude for both cattle and their herders to exist happily, the high pasturages offered rich grazing, and simultaneously allowed the home pastures, exhausted by the winter's feeding, the opportunity to recover. Moreover, in an era before the enclosure of fields, moving the herds away from the low-altitude farmland ensured that growing crops were safe from any risk of being trampled or consumed by the free-ranging animals. After a summer of rich, high-altitude grazing, the cattle – with the exception of a few vital breeding animals – would be taken along the high roads to be sold and butchered. Once again, the mountains were not a problem but a solution, even an opportunity. They enabled farmers to strike the balance between agriculture (growing crops) and pastoralism (raising animals for sale) that was essential for their survival and subsistence.

Thus, on any given day – especially in summer and autumn – in early modern Europe you would have found the mountains to be full of a gentle hum of motion of both humans and animals. Of villagers tending to the passes, or of travellers and traders crossing them. Of cattle or corpses being carried over the higher paths, away from the disturbance of towns. Of herds and even whole communities making their way up from the valleys where they had sheltered for the winter into the heights where they would live, work and – as I saw at Le Clou – worship with the sight of long mountain vistas beneath their feet.

Many of these mountain paths are still in use today; or perhaps it would be more apt to say that many of the paths trod by the feet of walkers, ramblers and climbers out for a day of recreation have a far longer history of life and labour in the mountains. Often without realising it, we may sit where coffins once rested, stumble over the almost-ruined remnants of an old shieling, and delight in reaching the top of a pass trodden by people for hundreds or even thousands of years before us. It is sheer arrogance to think we invented the practice of crossing, going into, or making ascents of mountain landscapes, or to fail to remember those who went before us. When we go on our mountain walks or climbs and revel in the sublime, empty 'wilderness', we are following in a crowd of premodern footsteps.

So, mountains were spaces of movement, as some of their inhabitants earnt their daily bread through fattening animals which they then led to market along the high paths. But that was not the only way to make a living from the mountains. Though they may have been relatively poor for growing crops, mountains were the habitat of wild animals which could be hunted and eaten, and the site of resources which could be mined and sold onwards. The works of a Dutch painter, Roelant Savery (1576–1639), vividly illustrate these activities. An oil painting of 1620 depicts a 'Wild Boar Hunt in a Rocky Landscape', with antlered deer in the middle foreground fleeing the scene as hunting dogs seize their prize. In his 1610 'Woodcutters in Mountains', men can be seen hard at work with saws and axes, though one woodcutter has bunked off his work to embrace a rather buxom young woman. Another of Savery's works, 'The Miners on the Mountain', survives as a printed engraving. For all their prominence in the title of the piece, the miners can only just be seen in the upper left-hand corner, carrying their burdens along a high mountain path surrounded by rocks and sheltered by trees.

In terms of mining, iron, coal and lead were all removed from mountains during the early modern period. We have already heard from Daniel Defoe – whether entirely accurately or not – on the condition of the lead miners of England's Peak District. The Flemish painter Lucas van Valckenborch (1535–1597) depicted, just a few years before his death, a vivid scene of a sixteenth-century mountain iron-mine (fig. 3). On the left-hand side of the oil painting, men work at a pulley system presumably designed to bring filled buckets of ore out of a vertical shaft. A rocky mass rises above them. On the right-hand side of the painting the encompassing crags create the outline of an hourglass. The bottom half of the hourglass is gloomy and dark: bridges across the deeper, open reaches of the mine can just be made out, with tiny figures crossing them cautiously. A plume of red fire and black smoke emanates from the smelting works. The top half of the hourglass is, by contrast, bright, opening out into a wide valley with blue-white peaks trailing into the distance. Valckenborch seems to be setting the beauty of the wider setting against the grim darkness of the mine.

3. Lucas van Valckenborch, 'Landscape with Ironworks' (1595), oil on canvas, 41 x 60cm. Madrid, Museo del Prado. (Photo: akg-images/Erich Lessing)

Wild animals and ore are both fairly 'obvious' resources offered by the mountains, but there is one more thing that, in our era of electricity and fridge-freezers, is all too easy to overlook: snow. Early modern Europe enjoyed a thriving 'snow trade', and where do you find snow, all year round? At altitude, on the sides of high mountains. Snow was used – alongside other preservation methods such as salting or curing – to keep food fresh, to cool drinks, even for treating illness. Snow also helped bring a taste of sweetness down the mountainside; records from Naples, in view of high mountains, note the serving of sorbet and ice cream at dining tables from the seventeenth century onwards. By the eighteenth century, the technology for making ice cream (a smaller vessel containing cream and sugar enveloped within a larger vessel containing ice and salt) had been made portable, so that small metal 'cones' of ice cream could be served to children on street corners, a predecessor to our modern-day ice cream vans.[21] Behind this rather charming vision lies more work for the mountain-dweller: labourers going up the mountainside, collecting and compacting snow, and then carrying it back down. Since ice, of course, melts, this would have been a regular activity – at least in the warmer summer months.

Has 'what happens in the mountains' really changed that much since the early modern period? The shielings, hafods and buailes are empty, and the production of ice cream no longer relies on workers making daily ascents for snow. However, mountains are still burrowed into – or, in the case of parts of the Appalachians, have their summits removed entirely – in search of valuable minerals. In the Scottish Highlands, hunting is still a regular activity, albeit often the preserve of wealthy visitors, some travelling across the world for the hoped-for thrill of shooting a twelve-point stag and taking his antlers home as a trophy. These visitors highlight one of the main ways that mountains make money today: through tourism. Visit any mountain town and you will likely find hotels, gift shops, places to purchase rucksacks, hiking poles and waterproof trousers, and adverts for organised mountain activities, whether guided hiking, skiing, cycling, or rock climbing.

As we saw in Chapter 1, mountain tourists are really nothing new – and they offered an excellent form of income to the real mountaineers. Josias Simler makes it clear how much travellers in the Alps in the sixteenth century relied on 'the mountain inhabitants' for their safety. Travellers afflicted by vertigo 'allow[ed] themselves to be led by the hand by local guides who are used to these heights' and were attached to their conductor by a safety rope when crossing crevasses. These crevasses were often concealed by deep snow and, in some regions, locals would plant poles to show the way. Many were more canny, as Simler reveals, for 'most of the time they neglect to do this, to force the travellers who don't know the route to hire their services'.[22]

Wherever you find a written account of an 'outsider' in a mountain landscape, you will likely find some reference to a local guide – even if they are almost always left in cloudy anonymity. When elite authors do say more than a few words about their guides – without whom they would probably have been limited in their ability to navigate the mountain landscape alone – they are often dismissive, 'othering' in many of the ways I traced earlier. John Ray (1627–1705), a botanist, was disgusted by the behaviour of the guide whom he had hired to 'conduct' him to the summit of Snowdon (the highest mountain in Wales at 1,085m). It was one of those days that those familiar with the Welsh mountains will recognise: rain falling so hard that the pair had no choice but to take shelter. They did later proceed to the foot of the mountain but, as Ray put it, his guide was so 'desponding' that he was 'forced to dismiss him', and to seek a new guide the next day, when it was cloudy but apparently not as rainy.[23]

I have to wonder what the interchange would have looked like from the point of view of that first guide. They were presumably fairly late to the base of Snowdon, having started the day in Caernarfon, over 10 miles away, and having been delayed by the rain. Quite possibly their gear (probably carried by said guide) was soaked through, and since it was September they did not have the long sunlight of an earlier summer day to rely on. I suspect this guide, familiar with the landscape and aware of how treacherous it could be after dark, pointed all of this out to Ray. But in the written record he was merely a 'desponding' employee, just as the Everest porters were lazy Orientals when they did not immediately leap to build stone shelters to sleep inside after a hard ascent carrying enormous loads. Whether the seventeenth century or the twentieth, the mountain incomer always – as far as they were concerned – knew best.

THE KNOW-HOW OF THE REAL MOUNTAINEERS

Despite what Ray and others assumed, the real mountaineers of early modern Europe were intimately familiar with their landscape and knew very well how and when – and when not – to navigate it in order to stay safe. Many of the techniques still used by climbers today date back to well before the birth of mountaineering as a leisure activity.

There are two main written sources that reveal most clearly what the mountain-dwellers of early modernity knew about their mountains. By knowing, what I really mean here is practical knowledge – where to build houses, how to traverse a glacier safely, and what to do if disaster struck and someone found themselves buried in an avalanche or slipped down a crevasse. Knowledge of this sort was recorded in a 1714 book by a Swiss theologian and historian named Abraham Ruchat (1680–1750). His *Les Délices de la Suisse*, or *The Delights of Switzerland*, was intended as a guide for foreigners, and it was clearly used as such by one William Windham (1717–1761), who wrote *An Account of the Glacieres or Ice Alps in Savoy* in 1741 and incorporated extensive translated passages from Ruchat's work in his own slim pamphlet. Windham ignored the advice of eighteenth-century inhabitants of Chamonix who insisted that to ascend any further into the glaciers would be 'very difficult and laborious'. He was, nevertheless, conscious enough of his own unfamiliarity with the mountain landscape to employ several of these doubting locals as guides.[74]

Much of what Ruchat wrote about, and what Windham dedicated substantial space to reproducing for his own readers, can also be found attested to in an earlier source. Josias Simler, whom I quoted earlier, was a friend and contemporary of Conrad Gessner, and taught everything from mathematics to theology at the Carolinum Zürich, the predecessor to the faculty for theology at the University of Zürich. He is known, according to his current Wikipedia entry, for being the 'author of the first book relating solely to the Alps'. I am generally suspicious of historical statements that ever claim someone or something as 'the first', since making such a claim at a conference full of historians is always likely to result in some early medievalist clearing their throat and saying, 'well, actually …' However, his *Vallesiæ Descriptio Libri Duo: De Alpibus Commentarius* (1574), or his *A Description of the Valais in Two Books, and a Commentary on the Alps* does contain a wealth of information about the region and its people as observed in his own time. These are interspersed with cross-references to descriptions of traversing the mountain environment as found in ancient authors. Taken together, these sources weave a vision of pre-modern mountain knowledge which far pre-dates the sport and apparently modern know-how of mountain climbing.

The section of Simler's book that I am especially interested in here is titled 'On the difficulties and dangers of Alpine journeys, and the manner in which they are overcome'. A mountain danger as prominent today as it was in the sixteenth century is that of avalanche, in which either a loose top layer of snow begins to tumble down the mountainside, gaining mass and velocity as it goes, or a slab of tightly packed snow goes plummeting down a slope upon the shifting or collapse of a weaker layer of underlying snow. Years ago, whilst on a winter mountaineering course in Scotland, I sat in the Clachaig Inn – a hostelry which claims as its icon a pair of ice axes – and listened to a presentation on avoiding avalanche dangers. The very next day we went and climbed Buachaille Etive Mòr (1,021m), conscious that only the week before two climbers had been killed on the same mountain by a slab avalanche. Even today, and even upon relatively 'little' mountains, avalanches are a serious business.

Both Simler and Ruchat note that an avalanche can be caused by sound; so similar is the list of aural triggers that one suspects Ruchat was familiar with the earlier Swiss text. In Ruchat's phrasing, 'A trifle will produce these terrible accidents … the flight of a bird, the leaping of a chamois, the firing [of] a pistol, a shout, speaking loud', and more.[25] Simler's version, listing 'the passage of a bird or other animal … or the cry of someone passing', elaborates that 'in the case of a cry, the air, agitated by the simple percussion of the voice … puts the snow into movement'.[26] Today, we would call this sound waves. In fact, loud noises do not cause avalanches, but we can hardly look down on Simler or Ruchat for their mistake, given that it is a myth which continues to circulate today.[27] I remember hearing a joke about the famously big-voiced actor, Brian Blessed, and his attempts to climb Mount Everest; the suggestion was that his companions would want to gag him anywhere near a snow slope to prevent just this unscientific disaster.

Despite this misconception, Simler's description of how an avalanche functions contains many accurate details:

> The snow forms first into a ball. Then after turning upon itself numerous times, this ball becomes irregularly enlarged. The uneven and massive form

of the ball hinders it from rolling further, and it slides downward with a greater violence, growing larger each step. Acquiring a considerable power, and enveloping in its fall rocks, trees, clusters of plants and animals, men, houses and everything it encounters in its path, the ball carries them all right down to the foot of the mountain.

He identified avalanches as being particularly likely to occur on steep slopes bereft of trees, when the warm sun of spring has softened the snow or in autumn and winter when new snow falls upon old. He even noted that there were two different types of avalanche, 'one made exclusively from fresh and soft snow ... the other which also drags old snow and carries with it a thick covering of earth'; the loose-snow and slab-type avalanches as recognised today. He recalled that the latter type of avalanche had, some years before, buried over sixty people crossing the Alps near the source of the Rhine.[28]

Such ominous memories meant that the local mountaineers took care to avoid them wherever possible. Simler observed that people would never build houses near slopes prone to avalanche, instead choosing sites with some barrier – a small intermediary hillock, a forested area – between the steep, risky slopes. Paths would sometimes, by necessity, have to cross high-risk areas, and the locals knew to make such a journey in the cool of the early morning, 'when the danger is least', and to move swiftly and quietly – remember the belief that loud noises caused avalanches.[29] Ruchat put it a shade more colourfully, stating that travellers through avalanche regions were encouraged to 'get through as fast as possible, as one would of out of a house on fire'. He also elaborated on the techniques for ensuring quiet when crossing with a herd of animals, which included stuffing cowbells with hay to temporarily prevent them ringing.[30] Simler's concluding comment on the matter of avalanche avoidance emphasised the importance of respecting local knowledge: it is 'the mountain inhabitants, who know these regions well and can guess the imminence of the peril by observing certain signs', to whom outsiders should turn for advice on when and how to travel.[31]

More than that, Alpine dwellers also knew what to do if travellers or neighbours were caught in an avalanche. One of the pieces of advice I was given at that training session in the Clachaig Inn was that if you did find yourself swept away by an avalanche, you should do your best to bring your hands and arms up around your face, in order to create a pocket of air essential for survival. Simler gave the exact same advice, noting that, 'if the unfortunate victim thus buried can … create a little space in front of his face, he can possibly breathe and perhaps stay alive there for two or even three days.' He was over-optimistic there, at least: today it is thought that the longest anyone has ever survived in an avalanche is twenty-two hours, and more conservatively your pocket of air is likely to win you two or three hours, not days, for help to find you. Fortunately for the sixteenth-century traveller, help *was* often on the way, as Simler recorded:

> … when masses of this kind of snow begin to fall, the mountain people ask themselves immediately if any travellers had set out that day, and calculating the elapsed time, they can guess where they were buried by the snow. In these cases, the most experienced among the rescuers digs right away and attempts to discover if anyone is buried there whom they can carry to safety and save his life.[32]

How, precisely, should you treat someone who has been buried in an avalanche? The early modern 'mountain people', as recorded by Ruchat, had the answer:

> When any one is found seemingly dead, without sense or motion, the first remedy is to plunge him in cold water. To some it will appear both barbarous and ridiculous to dip a man, who is frozen, and almost dead with cold, into cold water; but let them know that it would be certain death to any one to give him heat suddenly when he is frozen. They begin therefore with dipping him into cold water … afterwards he is put into lukewarm water, then proceeding by degrees, they get him into a bed well warmed.[33]

This guidance is spot on. The recognition that an avalanche victim may appear beyond help whilst still being alive is one that holds true to this day. Mountain rescue teams follow the adage of not deeming someone pulled out of an avalanche to be dead until they are 'warm and dead'. It is also, as Ruchat emphasised, extremely important to gradually warm up a person suffering from hypothermia, or even just extreme cold.

During the pandemic, I took up the perplexingly popular sport of cold-water swimming without a wetsuit. I have become, by necessity, an expert on the process of warming up a cold body and have observed in myself that the worst thing to do immediately after immersion in the North Sea is to jump straight into a hot shower. If I do, I begin to shiver and feel dizzy. This is because when the human body becomes very cold, it conserves as much core heat through 'peripheral vasoconstriction': the blood vessels below your skin become narrower, reducing blood flow. If you were then to get into a hot bath or stand in front of a blazing fire, your peripheral blood vessels would dilate, the warmth previous conserved in your core would be lost through your skin, and your actual body temperature would drop further. Your blood pressure (heightened by the cold) would drop rapidly, and if you were cold enough to start with you could go into shock or even heart failure. The words vasoconstriction, vasodilation and blood pressure would have meant little to the Alpine dwellers of early modern Europe, but they had spent long enough in the mountains to understand how the human body responded to, and could best be saved from, extreme cold.

Avalanches were not the only emergency with which the real mountaineers had to contend. Ruchat recorded a 'wonderful Adventure which happened some years ago' to a chamois hunter named Stoëri. He, along with two fellow hunters, was crossing a glacier when he fell into a crevasse which had been covered by recently fallen snow. He apparently fell so far that he was quite out of sight, for his friends thought it quite likely that he had been killed in the fall, but they still decided to mount a rescue attempt so that 'they might not reproach themselves with letting him perish, without endeavouring to help him', and ran to a nearby cottage to find a rope. They found no such thing, and were instead forced to cut up an old blanket into strips and tie it together to create a makeshift cord. Meanwhile, poor Stoëri was half-submerged in a deep pool of icy water. His feet didn't touch the bottom, and he kept his face and upper body out of the water by spread-eagling his arms and legs and wedging himself between the sides of the crevasse.

This is as exciting a story of survival-against-the-odds as can be found in the mountaineering sections of modern-day bookshops. Stoëri was resigning himself to death and 'recommending his soul to God' when help arrived at last, and his companions lowered down their home-made rope. He tied it around himself, and they had almost succeeded in pulling him out of the crevasse when

– as with all good survival stories – disaster struck again. The rope broke, or more likely one of the knots slipped, and he fell down once more into the icy water below. Half of the rope had come with him, leaving a length with his companions insufficient to reach him, and, worse still, in this fall he had broken his arm. Like the best mountain rescuers, however, they were not going to stop now. They cut the strips of the blanket in half again to gain enough length to reach him, he held himself out of the water with his good arm whilst tying the end round himself with his broken one, and they pulled him at last out of the crevasse, where he promptly fainted. They carried him to a house, and he made a full recovery.[34]

Knowing the mountains is not (and was not) just about knowing what to do in the worst-case scenarios. It also has a lot to do with knowing what equipment to use, and how to use it to tackle the challenges of the mountain terrain that those living at altitude would encounter on a daily basis. Both Ruchat and Simler made note of the gear that early modern mountaineers used to help navigate their local landscape – much of it would not be out of place in the rucksack of a twenty-first century climber.

Crevasses – like the one that poor Stoëri fell into – criss-crossed the glaciers which visitors, hunters and herdsmen alike needed to get over. In order to cross them safely, many carried boards or beams that they could temporarily lay across the gulf. This is exactly what is done to this day to help climbers ascend the infamous Khumbu Icefall on Mount Everest, except rather than wooden boards the guides set up 'bridges' of stainless-steel ladders across the deep gaps. Where modern climbers carry ice axes, which they can use both to support their progress up steep slopes and to probe thick snow for hidden crevasses, the shepherds of the early modern Alps carried long alpenstocks, described by Simler as 'batons with an iron point', and which Ruchat described as being used to check for treacherous gaps in the ice.[35]

Ice and snow, even when not forming and concealing deep clefts for travellers to fall down, offer challenges to basic movement. Ice is slippery and snow, when fresh, makes a misery of forward movement, as each footstep sinks deep into the new snowfall. In coping with ice, Simler commented that 'the travellers, shepherds, and hunters who habitually haunt the high mountains' employed various means to avoid slipping. Along with the alpenstock, they wore 'iron soles on their shoes similar to horseshoes, equipped with three sharp points', or used 'straps to attach spurs to the bottom of their shoes'. Ruchat referred to mountain travellers attaching 'iron cramps to ones shoes'.[36] The crampon, of course, is still a standard piece of equipment for any climber heading above the snow line. For deep snow, early modern mountaineers would take 'circles of wood ... and interlace them in every direction in the form of a trellis', and attach them to their shoes. As Simler explained, these snowshoes served to 'enlarge the imprint of the step, and they avoid being swallowed by the snow and don't flounder around'.[37] When crossing a blazing-white glacier or snowfield your average sixteenth-century chamois hunter would also wear 'a dark vizor in front of the eyes': snow goggles.[38]

With the exception of snowshoes, modern versions of almost every item listed above made their way into my rucksack for the week of winter mountaineering which saw me seated in the Clachaig Inn learning about avalanches. During the week's training, I also learnt about different techniques for safely getting down a mountain. One of our climbs saw us descending Ben Nevis by one of its steep gullies. It was slow going and exhausting, one foot after another into the deep snow, the slope so steep that I was more or less sitting on my bottom each time I took a step downwards. This was the type of challenging slope that Simler, hundreds of years before, described as 'steep and practically perpendicular'. Facing such a slope, his early modern mountaineers would cut a tree branch and then 'sit down and slide down as if they were on a horse'.[39] I had no such branch, but at our climbing leader's encouragement sat on my Goretex-covered bottom, held my ice axe firmly beside my right leg for self-arrest before the snow slope turned into a mass of rocks, and glissaded down the side of Scotland's highest mountain.

I was also taught how to use the snow to create belay points for lowering a weight – in modern mountaineering, this is generally another person – down a snow slope. In the sixteenth century, mountains played host not just to climbers on be-crampotted foot but to travellers with far more bulky entourages. Sometimes, the use of 'capstans and tackle' was required to lower heavy carts from the top of a mountain pass. Simler records with relish that one medieval chronicler even noted the use of such measures to lower horses when a royal entourage, headed by the Holy Roman Emperor Henri IV (1050–1106), crossed the Alps in midwinter in the eleventh century.[40]

The process was even, Simler also commented, 'known to the Romans'. Such an aside is typical of Simler, who whilst describing contemporary, sixteenth-century practices as observed in his own Alpine region, constantly had one eye on his massive mental library of classical references. He pointed out that Silius Italicus (*c.* AD 26– *c.*101) had written of avalanches in the context of the warrior Hannibal triggering one deliberately to bury his enemies, whilst Claudian (*c.* AD 370–*c.*404) and Xenophon (*c.*430–355 BC) had written of the dangers of the cold in the mountains.[41] Simler saw little need to separate the observations of writers from up to almost 2,000 years earlier from the knowledge and behaviour of the fifteenth-century Swiss mountaineers about whom he was writing. Archaeological evidence shows us that the equipment used to navigate the mountains in the sixteenth century dates back even earlier: in 2016, a rudimentary snowshoe dating to between 3800 and 3900 BC was discovered in the Dolomites and it is thought they were first invented even before that.

The people who lived and worked in the mountains of Europe in the early modern period were more than capable of navigating the challenges presented by their environment. They drew on long traditions of knowledge which, even if it lacked the formal scientific terminology of avalanche mechanics and biology available today, certainly enabled them to preserve and save lives. The sport of modern mountaineering did not invent crampons, avalanche avoidance, crevasse rescue techniques, or glissading. Such know-how belonged to generations of real mountaineers long before them, won through long familiarity not just with the lower grassy slopes of mountains but with their glaciers and snow-covered upper reaches.

EXPERIENCING THE MOUNTAINS – HISTORY ON THIN ICE

The third and final question I asked at the start of this chapter is by far the hardest to answer. What did the mountain inhabitants of early modern Europe think and feel about the mountains? The people whose thoughts and feelings I am interested in here did not leave any accounts written by their own hands, because they mostly could not write. Even if they could write, anything which they did record on cheap scraps of paper was far less likely to survive than the published writings of the educated elites, bound in sturdy leather and preserved in multiple copies in private and public libraries across the continent.

It is necessary, then, to put together a piecemeal story from a variety of sources. This includes reading between the lines of elite accounts, to see whether we can reach beyond their preconceptions to get into the minds of the ordinary people who sometimes appeared within their pages. Next, we might turn to archaeology: what physical remnants of human activity can we find in the mountains? Demographic data – bureaucratic documents such as censuses or church record books – can give hints about the movement of people through the landscape. Finally, an ethnographic approach involves looking at accounts of mountain experience from more recent communities living in similar ways to the early modern mountaineers we are interested in, and cautiously extrapolating from there. This is how a lot of what we 'know' about prehistoric lifestyles has come about: through looking at, for example, the behaviour of 'primitive' tribes existing in the modern day and living in a similar fashion to what archaeology tells us prehistoric man and woman did.

In each of these cases, an effort of interpretation and even imagination is required to make the step from the raw data – the remains of a church, the number of people who travelled up to a high pasturage – to hypothesising about what long-dead mountain-dwellers felt about the landscape. However, as I wrote in Chapter 1 regarding the 'truth' of any historical source, there is very little in history writing that is absolutely, scientifically verifiable. History is a mass of interpretations, of conclusions drawn by fallible humans from the available data. What follows are my interpretations and imaginings of what the early modern mountaineer likely thought about his or her mountains.

One elite account that might give us a sense, in a roundabout way, of how early modern mountaineers viewed the heights is Johann Jakob Scheuchzer's (1672–1733) account of Alpine dragons. The existence of dragons in the mountains were, he claimed, reported to him by many 'honest men' living in the Alps during his journeys through the region between 1702 and 1704. His 1723 *Ouresiphoites Helveticus* (which translates as 'the mountain-haunting Swiss man') included a series of engravings depicting their encounters. At the risk of unfairly critiquing the artist, it has to be noted that the draconic illustrations in Scheuchzer are more amusing than terrifying. In one, a heavily feathered snake with a bifurcated tail stares out of the frame, its tongue sticking out and its

4a & 4b. Illustrations from Johann Jakob Scheuchzer's *Ouresiphoites Helveticus* (1723), vol. 2. (Reproduced with kind permission from the University of St Andrews Special Collections)

pupils dilated in a comical expression of surprise (fig. 4a). In another (fig. 4b), a man with an axe over one shoulder raises his hand to his mouth as he encounters a dragon standing on its tail. The dragon's legs are also raised, but the overall mood evoked is less one of menace and more one of mutual startlement.

Scheuchzer – though he found the accounts of the dragons fascinating – was fairly sceptical about their actual existence. He concluded that bones reported to have been from a dragon were clearly those of a bear, and commented that it made sense that the locals often reported the appearance of 'dragons' after heavy storms, because such weather generally dislodged snakes and lizards (who might, in the telling, grow into dragons) out of their nests. He also suggested that they might have been allegorical, noting that the locals also referred to mountain rivers and waterfalls as 'dragons', and that when a flash flood occurred they would say that 'the dragon became unchained'.

In modern analyses, Scheuchzer's dragons are often held up as a symptom of early modern superstition and as a sign that those who lived near the mountains at that time tended to fear and avoid the heights where possible. I prefer an interpretation a little closer to Scheuchzer's own comments regarding waterfalls, which is that reports of dragons did not signify a baseless, irrational, or overwhelming fear of the Alpine landscape. As Simler and Ruchat made clear, mountains were dangerous places, and I believe that the 'false stories' of dragons were one of the ways the people of the Alps expressed the various dangers they might meet at high altitude, whether floods, crevasses, or avalanches. Crucially, the mountaineers of early modern Europe did not avoid those 'dragons'; they just made sure they had the alpenstocks and ropes to hand to deal with them.

In the interlude preceding this chapter, I stumbled upon the physical remnants of the alpage at Le Clou where cattle would be fattened up over the summer. Archaeological, demographic and ethnographic sources can be brought together to give us some insight into what life was like on farms like this one. After all, the cattle did not go up the mountains on their own.

Transhumance was practised across Europe. Jesper Larsson, a Swedish researcher in Agrarian History, conducted a detailed study of 'Labor division in an upland economy', asking, essentially, who did *what* when herds moved to the summer farms in seventeenth-century Sweden. The study focused on a single parish, Orsa Parish near Uppsala, and the records kept by the pastor during the late 1600s. These records included a rare 'herder register', which listed all the individuals who worked herding the cattle in the parish. This is an example of what I mean by demographic evidence.

What the register tells us is somewhat surprising, at least when set alongside our own preconceptions about the past. If I asked you to imagine an early modern cattle-herder, what would you see in your mind's eye? I suspect the figure you 'see' would be male, probably tall and strong. However, as the Orsa Parish register shows, the vast majority of herders were female. The Royal Superior Court of Sweden had, in fact, issued an ordinance in 1686 stating that boys should not work as herders (due, apparently, to fears of bestiality), and that clergymen should remind their parishioners of the preferability of appointing women to serve the summer farms. This is why Pastor Elfvius of Orsa Parish made his herder register in the first place. The register covered the years from 1687 to 1692, recording the names of 1,340 individual herders – over 200 per year – who made their way to one of over twenty summer farms in the parish.

Generally, each household sending animals to the summer farm appointed one family member to go and stay with the cattle, unless they were sending a very small number and were able to pool their herd with a neighbour, and thus only send one herder for multiple households. Households sent daughters, wives, or even maids to take responsibility for the herds they were sending upland. In 1687, six households sent boys and men to the summer farms, but the following year they were replaced with women. For every other year, as far as the register shows, the summer farms were exclusively inhabited by girls and women ranging in age from 8 to 82. The average age of inhabitants across the summer farms was 34.[42]

The herder register gives us a lot of data, but not much in the way of detail or emotion. Larsson theorised that the farms were an important space for young women to learn the basic skills of animal husbandry, and that the role which women served as the independent managers of the summer farms increased their status within the household when they returned to lower altitudes. As for what those girls and women felt as spring turned into summer and their transition out of the household and up to the heights, the best we can do is imagine. Did the young girls look forward to learning to churn milk into butter and cheese (since the only way to bring milk products produced by the milch herds back down the mountain to the home farms was to transform them into solid wheels), or did they dread being bossed around by the older women? Did the women leaving behind husbands and at least some of their brood of children breathe a sigh of relief at the temporary change in domestic responsibilities? When the work of day was done, did the women at the summer farms sit together and speak of all the things that being isolated from the opposite sex enabled them to speak of – of menarche, of lovemaking, of childbirth? As well as hard labour, was there laughter, teasing, secrets shared?

Ethnographic evidence relating to Scottish shielings would certainly suggest that the movement to and from the summer pastures was associated with celebration and sociability. By 'ethnographic evidence', I mean oral histories: the writing down of memories related by the last generations of people to use the shielings. Some of these recollections date back to the early nineteenth century, as antiquarians in the latter half of the century became increasingly interested in the daily lives of ordinary people. The best the historian of earlier eras can do is hypothesise that, perhaps, these memories of later shieling experiences represent the tail-end of older traditions dating back to the peak of shieling use in the 1700s.

In the Highlands, the act of moving from the home farm to the shielings was called a 'flitting'. One oral account, dating from the early nineteenth century, speaks of a 'small flitting' and a 'big flitting'. In April or May, a few members of the community would herd the 'yeld' beasts – those not yet bearing young or producing milk – up to the shielings, and would stay there to make any repairs necessary to the living quarters. In June, the 'big flitting' would see the milch herds led up to the summer pastures. In some cases, it seems that the shielings were treated as 'summer townships', with almost the entire community making the move and leaving relatively few people to manage the home farms – the opposite of the Swedish model. Sheep, goats, pigs and chickens were taken up, as well as cows. In the early eighteenth century, it was proposed that itinerant teachers should travel between shielings to ensure the ongoing education of the youngsters living and working at them.⁴³ The shieling-dwellers had a variety of tasks to fulfil. Herders were responsible for taking cattle away from the shieling buildings in the morning, in search of untouched pasture, and back in the evening. Poindlers had the unenviable task of rounding up the inevitable stray. Dairymaids made cheese and butter and, if the home farm were close enough, would carry milk back down the hill on their heads. The shielings were inhabited for between six and twelve weeks a year.⁴⁴

Around Loch Lomond, the shielings were generally inhabited by women and children, whilst the men remained at home to tend the fields. When they moved, they would not only take with them milking stools and equipment for churning butter but also spinning wheels and wool for making clothes and blankets. Whilst they stayed at the shieling women would take the time to traverse the mountainsides to collect wild plants, for medicines, and lichens for dyeing wool. Oral histories from the Outer Hebrides depict the weeks spent at the shielings as happy and sociable, with evenings spent playing music and dancing. The day of the 'big flitting' out to the shielings was one of excited energy; the moment when the men came back to help the women and children make the move home one of sadness, but also an occasion for one last celebration. Any remaining food was eaten, and everyone danced and sang the night away.⁴⁵ It is quite possible that these sorts of accounts are pretty heavily tinged by nostalgia, not least because the most recent oral accounts come from people who experienced the last years of shieling life as children. However, archaeological evidence of the remnants of whisky stills found near shieling sites would certainly point towards the idea that the time at the high summer pastures was not exactly one of all work and no play.⁴⁶

You may well be wondering, at the end of this messy collection of things the real mountaineers did, knew, and possibly felt, in the mountains, why any of this is important. Does it really matter if the people who visited the Scottish shielings for six weeks a year filled the mountainsides with music, or that people living in the Swiss Alps knew how to cross a glacier safely? Compared to the deliberate, impassioned, seeking-out of mountains by the modern mountaineer, this all seems fairly incidental: things which merely happened to have taken place on or around mountains. The mountain peasants described by Simler did not choose to learn about the mountain landscape, they were simply required to by the accident of being born amidst them. The drover taking his (or her?) herds to market did not go by the high ways because he thought they were beautiful, but because they were most convenient.

This is all true. It is also a crucial and fascinating difference between the mainstream modern experience of mountains and a far older way of experiencing them. Today, the most culturally prominent view of mountains is that of the outsider, going *to* and *up* them. Mountains are set apart, they are wildernesses to be protected, spaces to be enjoyed for leisure, for adventure. As such, we consciously celebrate them as special and as awe-inspiring.

The people discussed in this chapter did not *go* to the mountains, they existed *in* them. It may well be that from time to time they reflected to each other in conversation on the beauty of a mountain sunrise, or of the enjoyment of reaching the top of a ridge and seeing the valley spread out below them. These conversations, however, do not survive, though that does not mean they did not occur. Even in a modern era of social media updates there is much that goes unrecorded, such as the moments of chaotic joy in a well-loved family home. Instead, what remains are the scarce remnants of what it meant to *be* in the mountains in early modernity: the equivalent of the lines marking a child's height on the door-frame, or the silly drawings or signatures made before wallpapering that are the traces people leave behind of their lives within a home.

The details of daily existence are not quite as exciting as narratives of adventure, but we should not ignore them. They are all we have left of the real mountaineers who spent their entire lives among the mountains.

Vantage Point: Healing the Blind

Consider two images, produced two centuries apart. The more recent (fig. 5), a line drawing, depicts an almost perfectly diagonal mountain slope running upwards from the lower right-hand corner to the upper left. Two figures are balanced precariously on this steep, icy slope, connected by a rope; they are climbers. The lower figure is almost hugging the mountain, the rope in one hand and an alpenstock in the other. The pose in which the lead climber has been frozen is a dynamic one, with an axe lifted high above his head in advance of a downward swing to either grab purchase to pull himself up or to cut a foothold into the ice above. The direction of travel is upwards: the summit is out of view, but every line of the image points to its existence and to the intention the mountaineers have of reaching it.

This image was embossed in gold foil upon the cover of the flagship publications of the Alpine Club in the decades following its inauguration in 1857. *Peaks, Passes, and Glaciers* collated the achievements of early Alpine Club members in 1859 and 1862, by which point it became apparent that the club required an annual publication to record the exploits of their growing membership. For the first eight decades of its existence the *Alpine Journal* bore the same embossed gold image. Today, new issues of the *Journal* are cased in glossy photographic covers, more often than not depicting a snowy mountain summit, but members can still have their *Journals* rebound in the traditional boards at their own expense if they so choose.[1]

Two hundred years before the foundation of the Alpine Club, sometime between 1655 and 1660, the French artist Philippe de Champaigne (1602–1674) painted an oil on canvas, a large piece at around 1m by 1.4m, titled *Christ Healing the Blind* (fig. 6). The painting is remarkable. Just like the *Alpine Journal* cover, it is distinguished by a striking diagonal; a line of trees, from the bottom right to the top left divide the foreground from the background. Below the trees a crowd of people wearing robes of various colours fill a sloping path. One figure, in pale pink robes with a blue cloak, draws the eye. He has dark hair, a dark beard, and his right arm is outstretched in an almost casual fashion. The faces nearest are turned towards him, for this figure is Jesus Christ. A few paces from him, huddling into the frame of the painting, are a pair of crouched supplicants, reaching out of a natural cave augmented by man-made structures: these are the blind. The background, and most of the canvas, is made up of the view these two figures will see at the very moment that Jesus restores their sight. Below the crowd is a lake or a river with bridges across it and city walls running down to its banks. A tall crag, topped with more buildings, dwarfs the bridges and escarpments. Further and higher still, a blue mountain ridge vanishes into the distance.

These two images show two things that mountains have symbolised at different points in history. With the foundation of the Alpine Club in the mid-nineteenth century they came to symbolise heroism, the victory of human effort and ingenuity over the natural landscape. In the seventeenth century, when Philippe de Champaigne lived and painted, the mountains – the first thing the blind men would see after their miraculous cure – symbolised something quite different: the greatness and beneficence of God.

5. Cover artwork on the first volume of the *Alpine Journal* (1863). (Photograph of author's own copy)

6. Philippe de Champaigne, *Christ Healing the Blind* (1655–60). (Zip Lexing/ Alamy Stock Photo)

Chapter Three

The Meanings of Mountains

As a historian of mountains, I have spent a great deal of time reading and thinking about what we mean by the term 'landscape'. This is something which geographers, archaeologists and historians have also spilt a lot of ink wondering about. This may seem absurd. It is obvious what a landscape is – it is that thing out there, the trees, rivers and mountains.

However, the idea that the landscape is something 'out there', separate from humans, is a faulty one. Even that most inhospitable of continents, Antarctica, is dotted with structures old and new: the blue metal spiders of the Halley Research Station on the Brunt Ice Shelf; the wood plank, seaweed-insulated memorial of Scott's Hut on Ross Island. Meanwhile, its ice, throughout, tells the ongoing story of man-made climate change. Early on, then, scholars working in the field of landscape studies recognised that 'landscape' was much more than a natural, untouched object separate from humans. In 1925, the geographer Carl Sauer (1889–1925) coined the term 'the cultural landscape' to refer to the ways in which societies and communities inscribed themselves upon the landscapes they inhabited through establishing settlements, constructing buildings, changing the courses of rivers and streams, and more.[1] Thirty years later, the historian W.G. Hoskins (1908–1992) took his readers on a journey through the 'palimpsest' of the English landscape, emphasising the visible traces of human development still to be found of every age from Saxon times to the present day.[2]

Since then, historians and geographers have gone even further. Probably the most important figure in this respect is Denis Cosgrove (1948–2008), who defined landscape as a 'way of seeing', emphasising that the external world was 'mediated through subjective human experience'.[3] What does that mean? Well, it means that for Cosgrove landscape should not be discussed just as something *out there*, but as something *in here* within ourselves. As human beings we don't just leave physical traces on the terrain around us, we also attach meanings to it inside our minds. Some such meanings are personal, based on individual memories and experiences: I will always associate Blencathra with my engagement. Other meanings exist in the minds of many people, and influence how we engage with certain landscapes. Everest does not care that it is the highest mountain in the world – it is humans who have deemed that important, bringing hundreds of climbers to its slopes each year.

These cultural meanings can in turn be used in literature, art and advertising to evoke a certain mood and to elicit a certain response. A painting of an icy mountain landscape might make us feel a sense of emptiness, of cold, but maybe also admiration or awe towards the beauties of nature. By the same token, the brochure of an adventure travel agency might add human figures to a similar landscape, evoking a thrill of excitement and temptation to join them. Mountains are even associated with brands and products that have nothing to do with climbing. A mountain ringed with stars might remind you of sitting in the cinema waiting for a Paramount movie to start. Toblerone chocolate bars proudly bore the silhouette of the Matterhorn on their packaging until 2023. This changed when production moved partly out of Switzerland – the image of the Matterhorn also being deemed a protected national symbol. Mountains have meanings, and they range from the obvious (of course a snowy mountain will evoke a sense of cold) to the more surprising.

This chapter will trace the meanings of mountains in the early modern period, when mountains were certainly also cold, but the first Toblerone had yet to be tasted. I will be focusing particularly on literature and art. When mountains appeared in poetry, what did they serve to symbolise or represent? When and why did mountains feature in the backgrounds of artworks? What were some of the most common ways to describe and depict mountains, and what does that tell us about how people felt about them?

Fit for a Prince

In 1587, the poet and soldier Thomas Churchyard published a slim volume entitled *The worthines* [worthiness] *of Wales*, which contained a series of poems reflecting on different regions of the country and its notable features, such as castles and rivers. One poem, towards the end of the book, is titled 'A discourse of Mountaynes'.[4] According to Churchyard, the mountains of Wales had much to recommend them. They seemed to have been made by nature as 'a platform' for wide, delightful views. The mountain-dwellers, though 'plain', were long-lived, lusty-hearted, laughed in one another's company and delighted in the rugged landscape within which they existed. This stood in contrast to the valleys below which were rife with discord, wealth and treason. It was only in the mountains, where luxury and softness were unknown, that true contentment and peace could be found.

Nestled within this piece of sixteenth-century social criticism is Churchyard's extended exploration of one thing that mountains symbolised in early modern literature. 'You may compare,' he says

> ... a King to mountain high,
> Whose princely power, can bide both brunt and shock
> Of bitter blast, or thunderbolt from sky,
> His fortress stands, upon so firm a rock.
> A Prince helps all, and doth so strongly sit,
> That none can harm, by fraud, by force nor wit.
> The weak must lean, where strength doth most remain,
> The mountain great, commands the little plain.
>
> A mountain is, a noble stately thing,
> Thrust full of stones, and rocks as hard as steel:
> A peerless place, compared unto a King.

So, in Churchyard, mountains served as a metaphor for the qualities one looked for in the ideal king or ruler: they were strong; they protected and helped their subjects; and ruled confidently over those below them.

Mountains were also presented as the ideal environment in which to develop the qualities of a good king, such as in *Theuerdank*, a remarkable illustrated poem first published in 1517.⁵ It is thought to have either been written by the Holy Roman Emperor Maximilian I (1459–1519), or by one of his courtiers under close supervision. It is a romance, telling the story of the adventures of a young prince, Theuerdank, in his path towards marrying the princess Ehrenreich. Theuerdank represents Maximilian himself, and Ehrenreich is Mary of Burgundy, whom Maximilian had married in 1577, only five years before her early death in a hunting accident. As an object, the printed *Theuerdank* was a beautiful, prestige item: a brand-new typeface was designed for it, it contained over 100 woodcut illustrations, which in many copies were colour painted by hand, and was printed on vellum. Copies of the first edition were gifted to German princes, dignitaries, and to close friends of the emperor. As a story, *Theuerdank* is a *bildungsroman*, of which modern-day examples include *His Dark Materials* and *Harry Potter*: a story of the chief character's formative years and physical and spiritual growth into a (generally admirable) adult. It is the story of how an adventurous young prince became fit for his bride and fit to rule. He did so amidst the rocks, crags and dangers of the mountains.

The epic opens with a great ruler, King Romreich, lying on his deathbed choosing a husband for his daughter Ehrenreich. On his death, the new queen promptly and dutifully sends for her intended bridegroom: the knight Theuerdank. Meanwhile, three 'captains' who had hoped to win Ehrenreich's hand for themselves plot to ensure the knight never makes it to his new bride. Their names, Fürwittig, Unfalo and Neidelhart play on the German words for impudence, misfortune and envy. To travel from his homeland to his bride, Theuerdank must cross three mountain passes, each of which is guarded by one of the eponymously treacherous captains.

Theuerdank takes almost ninety chapters, each with its own detailed woodcut illustration, to overcome the challenges they throw in his path. Mountains, near and far, are a constant feature of the landscape depicted in the illustrations. Each captain pretends to be a friend to Theuerdank, and a loyal servant to the queen, before leading him into situations designed – unsuccessfully – to result in his injury or death. Many of these take place in the mountains.

Theuerdank's adventures with Fürwittig largely involve encounters with wild mountain animals. He kills a bear, slays several wild boars and goes on numerous chamois hunts. On one such hunt, Fürwittig hopes the knight will get his foot stuck in a rocky crevice, but he is rescued by a fellow hunter. On another he is led on to a narrow ridge where he must throw a spear whilst balancing on only one foot – a feat which he achieves, much to the frustration and false praise of the treacherous captain. On another chamois hunt, his climbing irons come loose, but this time 'God eternal' comes to his aid. Finally, the penny drops for the accident-beset knight and he knocks the treacherous Fürwittig down with a single punch.

The second captain, Unfalo, places Theuerdank in the path of yet more bears and yet more boars. He also instructs servants and pays peasants to engineer avalanches and rockfalls intended, in vain, to kill the knight. In one episode set in the high mountains, Unfalo urges Theuerdank to pole vault from one peak to another in order to impress the women. Theuerdank is on the verge of following Unfalo's suggestion when a fellow chamois hunter calls out to warn him against such a jump. Later, Theuerdank does manage to use a long pole – a critical piece of equipment in a chamois hunt – to swing himself down from a dangerous, windy ledge (fig. 7).

Being a fellow hunter or member of Theuerdank's entourage was not the safest place to be. Whilst Unfalo's attempts to harm the knight were largely unsuccessful, the dangers intended for Theuerdank frequently met their mark in the form of those accompanying him. In one episode, Theuerdank is armed with a sabotaged crossbow. When he attempts to shoot a passing game bird, part of the bow snaps off entirely, Unfalo's goal being that it would strike the hero. Instead, it knocks off his cap, and hits the man unwisely standing behind him. Twice, rockslides intended for him knock his attendants off their feet. Churchyard would later write that 'A prince helps all, and doth so strongly sit / That none can harm, by fraud, by force, by wit'. So it was that in each of these cases, as well as saving himself, Theuerdank also made sure to rescue those brought into danger around him.

Despite being led on enough dangerous mountain paths to last a lifetime (twice during this section Theuerdank slips and has to hang on to a mountain 'shrub' to save his own skin), it is only after being almost set on fire that the knight realises Unfalo does not have his best interests at heart and parts ways with him. His further 'adventures' with Neidelhart, the third captain, are more martial than mountainous, with numerous tests of his skills in combat. In the ninety-eighth chapter, Theuerdank finally meets Queen Ehrenreich. Foolishly enough he had allowed each of the three captains to live even after realising they had enthusiastically been trying to murder him, so they engineer one

7. Theuerdank pole-vaulting between mountain tops. (CC-BY, reproduction kindly provided by the National Library of Scotland)

last challenge for him: a tournament, suggested by allies of theirs and filled with knights from Neidelhart's friends and family. Theuerdank unsurprisingly vanquishes all, and the three captains are finally executed.

One would think that at this point the knight's trials would be over, but the queen – demonstrating that she was not a woman to be impressed by a man pole-vaulting in the mountains for the sake of it – points out that whilst he has shown himself to be valorous, all his deeds to this point have simply enlarged his own 'worldly honour'. Before consummating their marriage, she would have him launch a crusade for spiritual glory. The penultimate woodcut depicts Theuerdank, on a golden-armoured horse, leading an army to the Holy Land; mountains loom over the city in the background. The final chapter, omitting any narrative of his actual crusading, sees Theuerdank standing atop fourteen swords, arranged in a circle resembling a spoked wheel. This is an allusion to the popular medieval image of the 'wheel of fortune', over which the knight has triumphed.

So, in order to win the girl (and the kingdom), Theuerdank has to prove himself in battle and in his dedication to God. He started out by proving himself in the mountains. He demonstrated familiarity with the skills and tools used by the 'real mountaineers' of his time, he nimbly avoided the dangers of rockfalls and avalanches, and aided and protected his companions. Mountains *are* dangerous places in the epic of *Theuerdank*, but positively so, for they gave the hero the opportunity to prove, and develop, his mettle.

Not all would-be heroes fared as well in the mountains of early modern literature as Theuerdank. From 1559 to 1610, various editions appeared of a collection of poems entitled *The Mirror for Magistrates*. The 'mirror' was intended to reflect the deeds and downfalls of famous and powerful people so that people in positions of authority – such as magistrates – could learn from them. New versions were brought out throughout the decades, with further characters added each time. The 1587 edition added the stories of Caligula, the Roman emperor who was assassinated after turning most of his attention to building luxurious palaces for himself, and of Thomas Wolsey, the English archbishop who for a time lived in the pocket of Henry VIII, but fell out of royal favour and was lucky only in dying of natural causes before he could be brought to trial on charges of treason. It also incorporated the tale of King Brennus, a quasi-legendary, quasi-historical ancient British king. Brennus met his downfall not in self-indulgence or in failing to arrange the annulment of an excessively amorous king. Instead, he met his end in the mountains.

In the *Mirror*, Brennus makes two mountain ascents. One is a crossing of the Alps, the other an attempt to seize Delphi, a rich shrine to the god Apollo on the slopes of Mount Parnassus. The crossing of the Alps was a heroic act, taken when Brennus was at the height of his fame across Europe. The poet (speaking with Brennus's voice) compares himself to Hercules and describes the terrain as being filled with 'Great mountains, craggy, high, that touch the skies'. The paths were narrow and the brave travellers beset by 'ice, snow, cold, clouds, rumbling storms'. However, there is also a hint of the ridiculous in this vision of the bold king and his hardy army crossing the heights: sometimes, Brennus reports, the path was such that he and his soldiers resorted to sliding 'on buttocks down another breach'.[6]

The concluding verses of the poem are located on 'Mount Parnassus fair', the home of the Delphic oracle, where Brennus meets his downfall. The mountain itself acts as a fortress so high that the inhabitants need fear no enemies or build walls for protection, since its rocky sides are stronger and steeper than anything yet built by human hands. The peak is described as a source of 'fear and wonder', causing men to 'stare and gaze' at such a 'stately high and mighty hill'. Equally delightful are the ancient structures thereupon, in particular 'Apollo's temple high to heaven [which] above the rest doth stand'.

Here, Brennus claims, men were so 'bewitched' by the oracle that they brought copious worldly riches to the temple of Apollo. Unfortunately, his own soldiers were themselves bewitched by the temptations of wine before launching their assault. Such a party – drunk, hungover or, as Brennus put it, 'brainsick' – was no match for the Greek soldiers at Delphi, who were not only sober but also knew the mountain. They threw rocks on the British force, and guarded the easiest routes up the mountain, so that Brennus's troops were forced to attempt to scale steep rock faces. The environment itself turned against them:

> The ground did shake, and rent, and tempests rise,
> The hailstones mighty fall, the thunders roar,
> The lightnings flashing dazzled all our eyes,
> The Britons from the assault were over bore [overborne].

It did not take long for Brennus to lie wounded and defeated; whereupon rather than be captured alive, he stabbed himself in the heart after urging the reader to tread a wiser route than he and avoid doing battle 'with Delphos men'.[7] In a sense, Brennus's mountain ventures epitomise the classic (or even classical) tale of hubris, or pride. Having proven himself on the heights once, he overreached himself, and was defeated by both the landscape itself and the people who knew it best. In the *Mirror for Magistrates*, mountains symbolise power but also mark the limits of human strength and endeavour.

TROPES OR TRUTH?

Looking back at the mountain poetry of the preceding centuries, the mountaineer and critic Leslie Stephen observed with satisfaction that the poetry of modern writers such as Coleridge and Byron revealed

> genuine mountains, so to speak, of flesh and blood, not mere theatrical properties constructed at second-hand from old poetical commonplaces.[8]

The 'genuine mountains' of modernity were thus set against those of earlier writers who, in Stephen's assessment, merely repeated old clichés. Decades after Stephen, the literary scholar Marjorie Hope Nicolson wrote that the mountains

> upon which modern poets have lavished their most extravagant rhetoric, were for centuries described — where they were described at all — at best in conventional and unexciting imagery, at worst in terms of distaste and repulsion.[9]

The implication in both of these passages is that the early modern authors who used 'conventional' or 'commonplace' tropes to describe mountains were not really demonstrating anything beyond their ability to imitate earlier writers. They certainly were not, according to Stephen and Nicolson, communicating their real feelings for mountains.

This criticism makes sense when you think about how we view writing in the modern day. It is high praise to say that a novel is 'original', different from anything that came before. By the same token, works that are 'derivative' attract scorn from some quarters, even if they are commercially successful (not another vampire love story, sigh the critics, whilst copies fly off the shelves). Within the modern university, students' work is run through 'plagiarism checkers', software guardians against the least acceptable form of unoriginality. At the other end of the scale, marking schemes for essays in humanities subjects often suggest that work deserving of the best mark will in some way be 'original'.

The thing is, originality is very much a modern invention – rather like mountaineering. I am actually not certain this is a coincidence. As I will discuss further in the final chapter, modern mountaineering is intimately tied up with the idea of being first on a summit. To be original is essentially to 'be the first' to say a certain thing a certain way. But until relatively recently, originality was not the key marker of literary genius, just as being first on a mountain top was not the pinnacle of all possible mountain activities.

Let's take some of the classics of premodern literature. 'I sing of arms and the man', opens the *Aeneid*, a work so famous you've probably heard that phrase before even if you've never read the epic poem itself. And what was the *Aeneid*, in modern terms? It was essentially fanfiction of the *Iliad* – at once a retelling and a continuation of the story of the fall of Troy, from the perspective of the losers, the Trojans, rather than the victorious Greeks.[10] If you have not excised all memories of high school Shakespeare lessons from your mind, you might also recall that Shakespeare, though widely praised as one of the finest writers in the English language, rarely made up a story on his own. For example, *Romeo and Juliet*, first printed in 1597, had been a previous life as *The Tragical History of Romeus and Juliet*, a poem published in 1562 by Arthur Brooke, itself based on an Italian novella by Matteo Bandello.

So writers have not always been judged by their 'originality'. In contrast, a good poet in the early modern period was someone who was aware of literary precedent, of the ways certain ideas were usually expressed and the stories that had been told before, and was able to skilfully take them and incorporate them into their own work. Discarding every poem or piece of writing that used an apparent trope or cliché to describe mountains (or anything else, for that matter), would leave us with very little material to consider.

Even more than this, I think it is a mistake to assume that an idea or phrase is meaningless just because it is frequently repeated. It seems to me that tropes, or stereotypes, or clichés, become those things because enough writers deemed them to be apt ways to put something which felt true to them into words. Then, once something becomes a cliché, it still influences the way people think and act, even though they are well aware these are not 'original' ideas. It may be over 200 years since Robert Burns declared that 'My luve is like a red red rose', but lovers and supermarkets alike continue to present them as tokens of romantic affection. So, in fact, I think we have a lot to learn from the *unoriginal* ways people wrote about mountains in the past.

Often working as a historian is like doing a giant jigsaw puzzle, without the box lid to show what the finished product should look like. Not to mention that the pieces in the box actually belong to several jigsaws, and you have to sift through to find the ones for the jigsaw you are making. Every now and then, however, you stumble upon the gift of a source that effectively acts as the box lid, giving you a bigger picture all in one go.

At one point in my research, I set out to discover – if I could – how mountains were most commonly described in early modern English literature, and I found a rare 'box lid'. It was published in 1657 by a Yorkshire-born schoolmaster, Joshua Poole (c.1615– c.1656) and was titled *The English Parnassus: Or, a Helpe to English Poesie*.

You may remember that William Lithgow looked upon Mount Parnassus as a metaphor for the life of the poor but inspired poet. In ancient mythology, Parnassus was the home of the Muses and by the early modern period it had become a byword for poetic quality. Poole's *English Parnassus*, then, was a guide for young poets. It opened with a poem of its own, addressed by Poole to the adolescent students whom he taught. He entreated his promising youths to 'Accept and use then this my book' and to 'aspire / Unto the mountains top' of literary achievement.[11]

On my sixteenth birthday, I was gifted a Collins' *Good Writing Guide*. It contained essays on grammar, a guide to the correct use of capitalisation, commas, hyphens and quotation marks, tips for 'improving your power of expression', and many lists: of commonly misspelt words; of 'useful but unfamiliar words'; and of foreign words and phrases likely to appear in English-language works. Poole did not spend any time on spelling and capitalisation; though I have standardised spelling in quotations throughout this book, consistency in either was far from a priority for early modern authors. He also offered no 'handy tips'. Instead, he shared many lists, and many *examples*, drawn from the authors and works which he deemed to represent the best in English literature. Some we would still recognise as household (or at least high school) names today: Chaucer, Shakespeare, Milton, Donne. Others have declined in fame over the centuries: George Chapman and his translations of the *Iliad* and the *Odyssey*; Michael Drayton; George Sandys. In an age before originality, the works of these men offered the most reliable guide to the ideal way of expressing things.

The *English Parnassus* consists of three main sections. The first is a guide to rhyming, with an alphabetical list of all of the monosyllables that aspiring poets could pair together in order to please the ear. The second is what might be termed a poetic dictionary, with headings made up of every noun a poet might find themselves using in their works. Under each heading is listed all of the most appropriate adjectives or epithets to be attached to the noun, based on Poole's study of the texts listed above. The third section is another list of words followed by example passages or phrases relating to them in existing literature. (As an aside, this third section is, in modern terms, a plagiarism-checker's nightmare; Poole occasionally indicated in the margin where a specific quote came from, but often did not, and examples from different authors are mashed together with no indication of where one ends and another begins.)

In the poetic dictionary, 'Hill' and especially 'Mountain' receive particularly detailed entries. The 'Mountain' entry is given in full below; try reading it out loud, rolling the words around in your tongue and considering the associations – positive and negative – that they evoke in your mind:

> Moss-thrummed, rocky, shady, cloud-headed, insolent, steep, ambitious, towering, aspiring, mossy, hoary, aged, steepy, surly, burly, lofty, tall, craggy, barren, stately, climbing, sky-kissing, sky-threatening, cloud-inwrapped, high-browed, shaggy, supercilious, air-invading, hanging, brambly, desert, uncouth, solitary, heaven-shouldering, leafy, resounding, rebounding, echoing, thorny, inhospitable, shady, cold, freezing, unfruitful, lovely, crump-shouldered, sky-braving, pathless, lovely, cloud-touching, star-brushing, bushy, ascending.[12]

What a list! I love the craggy sound of a 'crump-shouldered mountain', or the sense of attitude evoked by the idea of a 'supercilious mountain'. And did you notice that 'lovely' evidently appeared so prominently in Poole's notes that he listed it twice? To me, this list highlights the real range of things that mountains could do in English poetry, the emotions and sensations that they could evoke. They could be towering, threatening, inhospitable, freezing. They could also be lovely, star-brushing, covered in leafy bushes and shouldering the sky.

In contrast to the lengthy entry in the poetic dictionary, 'Mountain' receives pretty short shrift in Poole's section of poetic exemplars. Under the heading, he gives just one line, with three phrases: 'The rocky ribs of the Earth. Earth's warts. Blisters'. This hardly seems laudatory – an easy example to prove that mountains were viewed with distaste in the early modern period, as blemishes upon the landscape. I think it is more accurate to say that reality is and always has been more complicated than unqualified admiration or unremitting denigration. Why shouldn't early modern poets have used mountains to evoke a whole range of different emotional responses?

In his list of poetic exemplars for the term 'Hill' Poole repeats the reference to mountains as warts. He also includes a very important list:

> Two rocky hills lift their proud tops on high.
> And make a vale beneath, Their lofty brows display.
> Earths dugs, warts, risings, tumours, blisters.
> Athos, Atlas, Hæmus, Rhodope, Ismarus, Eryx, Cithæra,
> Taurus, Caucasus, Alps, Appenine, Oeta, Tmolus, Ætna.
> Parnassus, Othrys, Cynthus, Mimas, Dyndimus, Mycale.
> Pelion, Pindus, Offa, Olympus, Helicon, Ida.

In the final four lines, which read almost like an exotic prayer, Joshua Poole reached through time and handed me the names of the mountains which loomed largest in the literary imagination of the seventeenth century.

Some of these mountains and mountain ranges receive their own entries detailing the literary precedents for describing them. Etna is (among other things) 'The vast Sicilian hill, whose jaws expire / Thick clouds of dust, and vomits flakes of fire'.[13] The Caucasus range is, fairly ominously, 'The Scythians snowy mountains on whose top / Prometheus growing liver feeds the Crop / Of Joves great bird'. It is also, rather more beautifully, the range 'Which with less distance looks at heaven by far / And with more large proportion shows the stars'.[14] Mount Ida, where classical mythology tells that Paris (of Trojan War fame) had to choose which of three naked goddesses was the most beautiful, is 'enchas'd with silver springs' and 'always crowned with plenteous crop'.[15] Parnassus, too, receives its own entry: 'The Muses forked hill. / With two tops reaching to the sky ...'.[16]

Mountains were clearly places of legend and mystery. Entries for various mythological creatures make it clear that they were home to more than just lions and strawberries. Satyrs are 'the goat-hair'd gods, that love the grassy mountains', and the most famous of them, Pan, is 'the mountain goat-footed God'.[17] The entry for 'Muses' highlights your likelihood of encountering them upon 'the forked hill / Of high Parnassus'.[18] If you were searching for 'the beauteous sylvan deities' (nymphs) you could expect to find them as they 'trip upon Mountains'.[19]

So, if you were one of Poole's poetical students and wanted to write about mythological creatures you might set your work in a mountainous setting. However, you could also use mountains to evoke, particularly, sensations or emotions. If you wanted to write about something cold, you might choose to say it was 'as cold as Alpine snow', or 'Cold as the top of snowy Algidus' (a mountain ridge near Rome).[20] Something violent could be illustrated with the image of lightning flashing against the 'surly brows' of 'proud mountains'.[21] Something or someone who was free could be described as being 'free as the mountain wind'.[22] If you wanted to talk about something that could never happen, you might mention seaweed appearing on the tops of mountains.[23] Conversely, if you wanted to capture the idea something that would last forever, you might call on the lovely image: 'While shades the mountains cast'.[24]

Unsurprisingly, 'white' could be described as being as white 'as mountain snow' or even 'like new fallen snow upon untrodden mountain' – which is a particularly striking image, since it suggests that the opposite concept, of a well-trodden mountain, was also present to the mind of the early modern author or reader.[25] Another unsurprising entry for mountains to appear under is 'high', and here Poole collects together a substantial compendium of past eloquence on the subject. You could say something tall was 'like Athos mount, or Erix steep and long / Or like old Apeninus raised on high'.[26] If you really wanted to capture a sense of height, you could go right to the top and refer to 'the Canarian Tenariffe', i.e. Mount Teide, then known as the highest mountain in the world, and upon which Marmaduke Rawdon in Chapter 1 climbed on the shoulders of his guide so that no one could be higher.

This section also contains some passages which I, at least, think are quite beautiful and evocative. Something could be described being as high 'as mountains on whose barren breast / The labouring clouds do often rest'. (This particular borrowed quotation of Poole's is still famous today, coming as it does from John Milton's *L'Allegro*.) Meanwhile, a high mountain is one 'which with less distance peeps into the stars', or which rises 'in stately height to parley with the skies'. Such peaks possess 'glittering tops, which fatal lightning fear', and 'invade the skies … and steal a kiss from heaven', and 'muffle up their heads within the clouds'.

Speaking of the mountain's barren breast, it is interesting to consider the instances where mountains appear as an appropriate simile or metaphor for another object. Waves, Poole suggests, could reasonably be described as 'liquid mountains' or 'moving mountains', and whales are 'the floating mountains of the sea'. But it is when describing breasts (something early modern poetry did with reasonable regularity) that mountains become particularly relevant as imagery. Breasts are like 'snowy mountains', 'warmer Alps', 'Venus Alps', 'swelling mounts of drive snow', or even 'Loves swelling Apennine'. Like mountains, they are divided by a 'valley', and produce fountains not of water but of milk.[27] The life-giving nature of both mountains *and* breasts loomed large in the early modern mind, and the equation between the two was made not only in poetry but also – as the next chapter will show – in scientific discussions about the origins of mountains as well.

I would like to end this exploration of Poole's *Parnassus* with one more example of how he thought poets should use the image of the mountain: in describing the first moments of the morning. Here, the image of the sun rising over a mountain landscape is recommended over and over again: 'Day sprung, and mountains shone with early beams / The morning night dismasks with welcome flame'.[28] You could describe the waking discovery of seeing 'Mountain tops new[ly] gilded over', or 'pearly dew sprinkling the mountain grass'.[29] Or perhaps you could write about the way 'jocund day / Stands tiptoe on the mountains top', or of how 'Aurora [dawn] now puts on her crimson blush. And with resplendent rays gilds over the tops / Of the aspiring hills'.[30] A mere poetic commonplace? Maybe. But it seems to me that its repetition tells us something important. Early modern writers recognised the pleasure of the sight of the sun rising over the mountains and expected their readers to appreciate it as well.

Putting mountains on the map

It may seem strange to include maps in a chapter about art and literature. Today, a map is not a piece of art; it is a functional, practical resource to help you find your way from A to B, or for learning about the geography of the world. Art and literature are also often imaginative, and we tend to take it for granted that maps represent a factual reality. Even today, though, maps are not 'correct' in all possible aspects. Any attempt to reduce the complexity and reality of the outside world into a form that can be pinned to a wall or accessed on a smartphone is going to involve leaving out some details, and even consciously distorting others.

Consider the Mercator projection of the world, upon which many people's mental vision of the shape and size of the world's continents is unconsciously based. The act of depicting in two dimensions – i.e. on a flat surface – the globe of the world is always going to come with distortions. The Mercator projection fits our spherical world into a neat rectangle. In so doing, it minimises the size of landmasses nearer the Equator and enlarges those close to the poles. Looking at a Mercator map of the world you would be forgiven for thinking that Greenland and Africa are more or less the same size, when in reality Greenland would happily fit inside the Democratic Republic of Congo, and is really fourteen times smaller than the entire African continent.

Even looking at mapping on a more detailed scale it is worth thinking about how the choices made by map-makers (or, today, app developers) shape the way we view and experience the world around us. The ubiquitous Google Maps is not a neutral record of the world outside of your door. In town, it highlights some shops before others. And as for out of town, well – let's just say that in the course of my attempts to put the travels of friends such as Thévenot and Olearius on the map I have had plenty of opportunities to tug at my hair in frustration at Google's lack of interest in labelling the natural landscape with the same energy it applies to making sure you know exactly where your nearest Pret a Manger is.

So even today maps distort reality just as much as poems or stories can. They tell us as much about what the map-maker thinks matters most as they do about the actual topography being depicted. This is even more true for the early modern period. Producing maps in the sixteenth and seventeenth centuries was a costly undertaking. The examples I am going to discuss were not intended as aids to route-finding. Instead, they were made to be admired in moments of leisure, and adorned the libraries of some of the wealthiest people in Europe.

For every printed page of an early modern atlas there had to exist a corresponding physical engraving – sometimes massive in size – and a press capable of transferring the inked engraving on to a page. Colour would often be added by hand. And this merely describes the final stages of the publication process; for areas for which plates (or maps upon which to base new ones) did not already exist, a surveyor would have to be engaged, and then a craftsperson capable of transforming their draft maps into engravings. A single sheet from the 'first' modern atlas, Abraham Ortelius' *Theatrum orbis terrarum* (the Theatre of the Orb of the World) will set you back over £4,000 today, and deservedly so. They truly are pieces of art.

The largest and most famous early modern atlas is Joan Blaeu's *Atlas Maior* (literally 'the bigger atlas'), published in Amsterdam between 1662 and 1672. It is a staggering thing just to look at on a screen, and I have been lucky enough to handle a physical copy a couple of times. It was published in Latin, French, Dutch and German, and the different editions range between nine and twelve volumes each. Each volume is a heavy tome, almost half a metre tall, and not a book you want to drop. It was the most expensive book ever produced in the seventeenth century. A skilled craftsperson on a good salary would have had to work for a year to afford it – not that it was really intended for them. In 2018, a copy sold at auction for over €500,000.

The *Atlas Maior* contains far more than just a series of bland, factual maps. On some pages, illustrations surround the maps depicting the type of people one could expect to meet around the world. Along the sides of a map of the Americas we find 'Mexicani' in feathered robes, the half-naked 'king and queen of Florida' and a pair of completely naked, remarkably well-muscled 'Brazilian warriors'. The seas around the mapped continents are filled not just with islands but with ships, well out of scale, and a variety of sea-monsters. Looking closer at the continents, one sees not just the names of settlements and straightforward topographical details, such as mountains, rivers and forests, but also small illustrations of people and animals; towards the southern tip of South America there stalks a creature that rather resembles a well-groomed poodle.

Most mountains in the *Atlas*, but not all, are depicted generically: little open-ended triangles or upside-down 'Ws' with shading to the edges. When you get to a larger scale of map, the mountains become more detailed – but not necessarily accurate in the sense of being perfectly to scale. One of my favourite examples of this is the map of Jan Mayen, a volcanic island 620 miles west of Norway in the Arctic Ocean. It was mapped and named by Dutch explorers in 1614, and from the sea it is dominated by the looming mass of the Beerenberg (2,227m). The map in Blaeu's *Atlas* is actually fairly accurate, insofar as it delineates the general shape of the island, but the Beerenberg is enormously exaggerated. The landscape is three-dimensional, as if looking down at the island from an angle rather than directly above, and the mountain's sweeping sides appear to be almost vertical, its pyramidal shape transformed into a sharp finger reaching up into the sky and looming far over the craggy hills depicted below it (fig. 8).

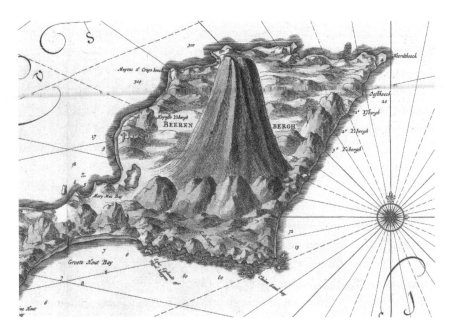

8. The Beerenberg, Jan Mayen Island, as depicted in the Blaeu *Atlas Maior*. (CC-BY, reproduction kindly provided by the National Library of Scotland)

The thing is, whether one considers this to be *accurate* or not depends on what you are expecting a map to communicate. You cannot read the height of the Beerenberg from this seventeenth-century map, and nor can you calculate the horizontal distance from its base to its summit. What this page of the *Atlas Maior* does communicate, very well, is the sensation that those Dutch sailors probably had as their boat neared Jan Mayen, and they looked up from their tiny deck to the mountain looming above them.

Maps could also communicate a sense of what was most culturally significant about a mountain, telling people what they should be thinking about in relation to a specific peak rather than just what they might be likely to see. One striking example of this in the *Atlas Maior* is the map of the Holy Land. The importance of some locations is indicated in text annotations, such as *habitatio Ismaelis*, the dwelling-place of Ishmael (the often-forgotten son of Abraham and his wife's maidservant, Hagar). More significant Biblical moments, however, are depicted graphically. Near the centre of the map, you can find Mount Sinai, depicted as a slightly larger, craggier mass than the

9. Mount Sinai as depicted in the *Atlas Maior*. (CC-BY, reproduction kindly provided by the National Library of Scotland)

'generic' mountain ranges alongside it (fig. 9). However, what really distinguishes it is the ring of clouds at the summit of the mountain and the tiny figure with its stick-like arms held up towards heaven and holding something rectangular in shape. These few lines on the engraving illustrate Moses receiving the Ten Commandments. At the foot of the mountain, more stick figures dance around a pillar, topped by a rotund shape with a neck and a head: the Israelites worshiping the golden calf in Moses' absence. Here, the map-makers were not trying to give their readers a sense of what Mount Sinai was really like topographically but rather to emphasise why it mattered. As we know from Jean de Thévenot's account of religious tourism in Chapter 1, these events were indeed the ones that came to travellers' minds when they visited the actual landscapes depicted in this portion of the *Atlas Maior*.

Another example of a 'mis'-represented mountain – in the sense of its depiction not really matching the visual reality – in the Blaeu *Atlas* is more local to me: Mormond Hill in Aberdeenshire, Scotland. In Gaelic, it is *A'Mhormhonadh*, meaning 'great hill' or 'great moor', and it is a towering 234m high. Relative to other peaks depicted in the *Atlas Maior*, it is a veritable pimple.

Mormond Hill – and all of Scotland – took a fairly roundabout journey to entering Blaeu's *Atlas*. At the end of the sixteenth century, a Scottish minister named Timothy Pont (*c.*1560–1627) took it upon himself to survey and map the kingdom of Scotland.[31] His dream was to produce a detailed, published atlas of the entire country, but he died shortly before the completion of his survey, and his remarkable maps remained in manuscript form only for many years to come. Enter the Blaeu publishing family. Long before the publication of the final *Atlas Maior*, Joan Blaeu and his father, Willem, had been putting out feelers for maps upon which to base their earlier atlas projects. They discovered Pont's maps, and engaged the cartographer Robert Gordon of Straloch (1580–1661) to re-draw and add to them. So it is that when tracing the depiction of Mormond Hill in early modern maps we can find it in Pont's original sketches, Gordon's later revisions, and the final published version in the *Atlas Maior*.

What fascinated me most as I set these different mappings alongside one another is the fact that Mormond Hill, on paper, literally grew over the course of the decades, from something which fairly closely resembled the view 'from

10a, b, c & d. Details from Timothy Pont's (a) and Robert Gordon's (b and c) hand-drawn maps of Scotland, and the printed version in the Blaeu *Atlas Maior* (d).

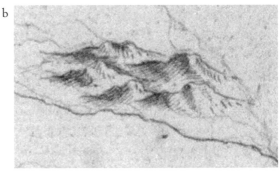

the ground', to something far larger and craggier. If you do an online image search for Mormond Hill, you'll be able to get a sense of its profile – a fairly gentle, undulating promontory amidst a relatively flat surrounding landscape. You'll also see a few features that Pont and Gordon never did: its hairy tuft of communications masts and satellite dishes; the figure of a white horse, cut into the hillside in the 1790s; and a white stag, cut in 1870.

Pont drew his map of Buchan – northern Aberdeenshire – sometime between 1583 and 1596, and his depiction of what he terms 'moir mont' is fairly true to life. You can see from Pont, just as you can from a modern OS map, that Mormond Hill is a miniature Parnassus – double-topped, linked by a low saddle (fig. 10a).

Robert Gordon's maps date from between 1636 and 1652. Gordon was called 'of Straloch', because this was the name of the estate he owned, a short distance north of Aberdeen and 30 miles south of Mormond Hill. Of his five maps taking in the region of Aberdeenshire every single one depicts the 'great moor'. One of these five is particularly striking, since it is really a detailed map of the west coast of Scotland with the easterly coastline of Aberdeenshire providing a contextual outline, with only a few towns and rivers adding detail. The eye is therefore irresistibly drawn to the pencil lines depicting a craggy mountain massif set just back from the coastline (fig. 10b). It is not labelled, but it is right where Mormond Hill should be, and there is no higher promontory in the area that it could correspond to. But the gentle, double-topped slopes of Pont's Mormond Hill have grown in stature, number and steepness.

In his two more 'zoomed-in' maps of the region, this effect is magnified. In one map, which seems to be partway between a sketch and a finished map, 'Moirmond Hill' is suddenly massive, with ten heavily shaded, steep tops reaching up towards the sky (fig. 10c). In what seems to be a more polished version of this map, the hill has fewer tops, but those which exist still look rather less friendly to the casual walker than Pont's original contours.

Finally, we come to the *Atlas Maior*. In the map of 'Aberdonia & Banfia', we find that the humble Mormond Hill has grown to almost Alpine proportions and profile. The highest point of the Mormond massif – as it is depicted here – is a sharp, high ridge, with nearly black shading communicating the sense of a stark cliff-face that would require rock-climbing techniques if you wanted to ascend it (fig. 10d). Viewed from a distance, it is the most eye-catching feature of the north-eastern portion of the map. A well-educated continental traveller

– like Adam Olearius from Chapter 1 – might well have had the chance to look at a copy of the *Atlas Maior*. Had Olearius done so, and had he then visited Scotland, he might have been disappointed to discover that the promised towering mountain was little more than a gentle moorland.

However, just as with the over-exaggerated Beerenberg, it misses the point to expect the *Atlas Maior* to be accurate for every single landscape it depicted. Again, the purpose of the *Atlas* was not to serve as a reliable road map. It brought maps of regions from around the world together, enabling readers to get a sense of places they would likely never visit, whether the Holy Land, islands in the Arctic Ocean, or Aberdeenshire. In its final, craggy iteration in the *Atlas Maior*, Mormond Hill holds its head up high alongside substantially taller mountains. If you look again at the actual landscape 'on the ground', this actually seems only fitting. Mormond Hill may be small, but it is the biggest geological feature for miles around – it catches the eye. In growing so substantially in order to fit into the *Atlas Maior*, the depictions of the Hill of Mormond do not tell the viewer exactly what it looks like, but they tell them something just as important if not more so: its visual prominence within its own local space. What made a mountain in the early modern period was not about absolute height. A pimple in the Alps could be a 'great hill' in the plains of Aberdeenshire.

When mountains speak for themselves

Blaeu's *Atlas Maior* was serious stuff: a prestige object that attracts as much awe and admiration today as it did when it was first printed. By contrast, the next source I want to talk about is one so quirky that it has provided the basis for an adult colouring book. This is Michael Drayton's *Poly-Olbion*, published in two volumes ten years apart, between 1612 and 1622. Drayton (1563–1631) was a prolific poet but has fallen into relative obscurity today. His literary legacy has probably suffered from the fact that he lived and wrote alongside writers of eclipsing fame. He was friends with William Shakespeare to the very last: the vicar of Stratford-upon-Avon claimed that it had been a 'merry meeting' (i.e. a drinking session) between Shakespeare, Ben Johnson and Drayton which had precipitated the bard's death.

Drayton's *Poly-Olbion* was something called a chorography. Looking into its etymology explains this distinctly early modern genre very well: the word comes from a pairing of two Greek words, *khōros*, meaning 'place', and *graphein*, meaning 'to write'. A chorography thus sets out to 'write a place'. One of the most famous early modern chorographies was William Camden's immense *Britannia*, which was first published in 1586 and described each county of Britain in detail, including not only its landscape and towns but also its history and any anecdotes of antiquarian interest.

Drayton's *Poly-Olbion* distinguished itself from its densely printed prose predecessors in two ways. Firstly, it was written as a poem, in rhyming couplets. Secondly, whilst previous chorographies of Britain did contain county maps, those in *Poly-Olbion* were a little bit different; they were light on cartographical detail, and heavy on fanciful human figures intended to represent the hills, woods and rivers described in the poems. It is these maps which formed the basis, in 2016, for *Albion's Glorious Ile: A Hand-Colouring Book from the Songs of Poly-Olbion*, when an academic project on Drayton took advantage of the trend for adult colouring books to bring Drayton's intriguing maps to a whole new audience.[32]

The poem takes the form of a 'muse' travelling around the British Isles, regularly handing the microphone – as it were – to various rivers, mountains, and even regions to speak for themselves. The 'song' for Yorkshire sees the three Ridings in a heated debate for seniority, with each taking great care to note the 'brave' mountains and fine rivers within their borders.[33] In Lancashire the River Ribble emphasises, somewhat controversially, her 'Birth' in Yorkshire, from the foot of Pen-y-Ghent. That mountain, her 'proud sire ... takes pleasure in my course', whilst she imagines other peaks gazing down with 'smiles' as she wends her way into Lancashire.[34]

Ribble is not the only landscape feature in *Poly-Olbion* to possess an inflated ego. One passage, titled an 'Ode to Skiddaw', caused me to pause when I first came across it. Skiddaw, at 931m, is one of the highest mountains in the Lake District, and the highest of the so-called Northern Fells. I have climbed Skiddaw several times in my life and have an ambivalent relationship with that peak. It is a lovely looking hill from below, with gently sloping ridges and a neatly rounded, conical top. However, the final ascent to the summit involves a slog up unpleasant, slippery scree and, whatever the weather in the valley, it has always been miserable up top every time I have visited. The first time, what had looked like an unremittingly sunny, calm summer's day at ground level suddenly turned into a hailstorm halfway up. Once we descended from the cloud, my older brother – perhaps a touch over-dramatically – panicked that his ungloved hands had become frost-bitten. A few years later, I was almost flattened on the descent from the summit by a mountain-biker speeding down the slopes. I have no memories of pleasant views or picnics on that hill, only wind, rain and near-decapitation by bicycle.

Drayton's Skiddaw, however, is smugly proud of himself. Having been introduced by the Muse, who highlights his double-topped resemblance to Parnassus, he declares that there is 'not a nook' into which he, from his 'glorious height', cannot pry. The 'great hills' around him are merely his attendants (he calls them 'pages', as if he were a bold knight like Theuerdank). It seems I was far from the first to note the poor weather on his summit, for he does admit that

> When my helm of clouds upon my head I take,
> At very sight thereof, immediately I make,
> Th'inhabitants about, tempestuous storms to fear,
> And for fair weather look, when as my top is clear.

Skiddaw has still more to say of himself. He peers around, spying the 'English Alps' (the Pennines), which 'look far off like clouds, shaped with embattled towers', and which view his distant height with envy. The River Derwent dances beneath him and just as he gazes down at her like 'some enamoured youth, being deeply struck in love', so too does his watery 'mistress' often look back up at him to confirm that his 'brave bi-cleft top, doth still her course pursue'. Being a mountain, Skiddaw has a long memory and recalls the Romans raising altars on his flanks. As far as he is concerned, the gods to whom these mountains were dedicated are also still present, and he ends by addressing them directly. With his characteristic confidence he is certain that 'Ye genii of these floods, these mountains, and these dales ... hold me Skiddaw still, the place of your delight.'[35]

For all that I cannot bring myself to agree that Skiddaw is a place of entirely unqualified delight, I love this passage, and others like it in the *Poly-Olbion*. It is so very unlike any modern way of thinking or writing about mountains. It is one thing to find passages delighting in the mountain landscape, but it is something else to find the mountains declaring their pride and delight in themselves. It is a less-reverent but somehow more personal way of thinking about the natural landscape, as characters with voices and

11. Detail from map of Cumberland and Westmorland in Michael Drayton's *Poly-Olbion* (1622). (CC BY-SA 4.0, Folger Shakespeare Library)

personalities rather than simply topographical features to paint, photograph, or ascend. And if we turn to the map for Cumberland and Westmorland in *Poly-Olbion*, we find Skiddaw depicted not merely as a mountain, but as a man, legs crossed insouciantly, a long stick in one hand, a large round hat upon his head, and his gaze turned outwards, enjoying the admiration coming his way from the similarly anthropomorphised mountains and rivers surrounding him (fig. 11).

Sacred and ancient: mountains in art

Earlier, I described the work of the historian as that of putting together a puzzle to make up a larger picture. The internet has precipitated a sea-change in historical research in terms of the range and number of puzzle pieces it is possible to collect.

Using online databases, you can draw together historical material existing in different physical collections across the world in a matter of seconds. What might once have taken close reading of hundreds of printed catalogues can now be achieved with a few quick clicks. The historian's work – of filtering the results, analysing them, and putting them together to create a coherent picture – can then begin.

When I came to explore the representation of mountains in early modern artworks, I benefited enormously from just this kind of twenty-first century shortcut. I started with a database called ArtStor, which contains more than 2 million images – of archaeological finds, of paintings, of modern architecture – all capable of being filtered through a wide variety of search terms and criteria. I set the date range to capture any images of artworks dating from 1500 to 1750, and then, over the course of several weeks, searched on anything titled or tagged 'mountain', 'mount', or 'mt'. This turned up a lot of irrelevant results (there being limitations to even the most efficient search engine) but every image that substantially incorporated a mountain in some way I copied into my research notes, along with full details about the medium, size and current location of the work. It was through this exercise that I discovered the works by Roelant Savery and Lucas van Valckenborch depicting the labour of woodcutters and miners on the mountains which featured in Chapter 2.

What else did I find? Well, I found a lot of mountains, which would have been surprising if I had still been operating under the assumption that no one liked mountains back then – why draw or paint something you found detestable to look at, and which your patrons would prefer not to think about? All in all, I collected around 300 images, which I divided into three broad categories: those which featured mountains as part of a religious scene; those which did so in illustrating a classical myth; and those which depicted mountains within naturalistic landscape paintings or drawings.

The majority of the images I collected fell into the final category – including the paintings of woodcutters and miners mentioned above. Quite a few engravings and etchings in particular chose to depict the sort of things which happened on mountains; as Theuerdank would lead us to expect, the hunting of the mountain goat attracted the attention of more than one artist, as did men herding their mules across mountain passes. Mountain landscapes were often drawn or painted as containing buildings or other human structures: castles, watermills, bridges, cottages, roads, inns, churches, entire towns. There were also people in these sixteenth- and seventeenth-century artistic mountains: hunters and herders, but others too. A lovely 1628 engraving by one Oliviero Gatti shows *A Young Man with a Telescope on the Summit of a Mountain*. Another painting by Roelant Savery, preserved by the Flemish engraver Aegidius Sadeler II (1568–1629), shows an *Artist Seated on a Rock*, seated by a river beneath rugged cliffs and with the suggestion of high mountains in the distance behind a rickety bridge. This was surely painted with a wink towards self-portraiture, given the number of times Savery turned his own hand to mountain landscapes.

What sense of or emotion for mountains do these naturalistic landscape depictions communicate? There is only one I could find that really suggested 'mountain gloom', and that is a dark and dramatic piece by Francisque Millet (1642–1679). Dating to about 1675, oil on a mid-sized canvas (around 1m by 1.25m), his *Mountain Landscape with Lightning* is full of brooding hills casting grim shadows, with the face of the peaks in the upper left-hand side of the painting obscured by dark clouds (fig. 12). A flash of lightning surrounded by almost completely black clouds, which look more like smoke, bisects the background of the painting, drawing the eye to the only portion of visual relief in the upper right-hand quadrant, where the viewer can peer through the clouds and the gloom towards a snowy ridge far in the distance. Figures in robes – not dissimilar from those featured in Philippe de Champaigne's *Christ Healing the Blind* – toil up a path in the foreground, looking out with alarm across the valley towards the lightning storm. These figures add a human element – a lens for the viewer of the painting to imagine seeing the scene through – to a painting otherwise dominated by a severe and frightening landscape.

This painting, so dark – with years of dust, I suspect, as well as the artist's choice of shades of paint – is the exception in terms of the mood in which mountains were generally pictured. The works of Joos de Momper (1564–1635) utilise a very different palette. His mountain landscapes are characterised by brownish foregrounds, usually some sort of mountain pass, usually with figures, buildings and trees, which serve to frame a wider view, dominated by blues and whites, of mountains – variously jaggy, craggy and rolling – receding into the distance (e.g. fig. 13). Momper's bright, shining peaks are far more characteristic of the standard visual and emotional palette of sixteenth- and seventeenth-century landscape paintings.

What did mountains 'do' in images of the landscape? They gave artists the opportunity to add a sense of wider space, distance and open air to their

The Meanings of Mountains

12. Francisque Millet, *Mountain Landscape with Lightning*, oil on canvas, 97.3 x 127.1cm (c. 1675). (National Gallery, Bought (Clarke Fund), 1945)

13. Joos de Momper, Große Gebirgslandschaft ('large mountain landscape'), 209 x 289cm (c. 1620). (Photo: akg-images)

paintings, to stand in contrast to the detail and oftentimes sense of enclosure offered by their foregrounds. However, mountains being 'in the background' did not necessarily mean they were a secondary detail; in many of the depictions discussed above it is the mountains which are the most eye-catching, with the foreground details leading the eye towards them. With the exception of Millet's lightning-struck landscape, it seems that artists regularly chose to depict mountains in order to promote a sense of enjoyment and appreciation in those who viewed their paintings of them.

The smallest but by no means least-interesting category which I discovered during my inventory of early modern artworks was that of mountains which featured in depictions of stories drawn from classical myth and literature. One example was the story of Psyche, a mortal woman so beautiful she was told she rivalled even the goddess Aphrodite. This statement naturally irritated said goddess, and as anyone even vaguely acquainted with Greek mythology should know, irritating divinities is not a good idea. Despite her beauty, no one asked to marry her, preferring to admire her from afar. Eventually, her father went to the Oracle of Delphi (located, as we know from the legend of Brennus, on the mountain Parnassus) for advice. Another god, Apollo, possessed the priestess of Delphi and informed Psyche's father that she would marry a terrible beast, and that he must dress her as if for her funeral and take her to the highest mountain in his kingdom. Psyche's story had a happy ending (she married the god Eros and was eventually made a goddess herself), but the moment of pathos of the young woman, head bowed as she sat on a litter carried by servants, a curling path behind her showing the upwards route to her doom, is one mountain image which exercised weavers and engravers of the early modern period alike.

Another prominent mountain myth is that of Atlas. Atlas was a Titan, one of the 'pre-Olympian gods' who came before the main cast of Zeus et al. that we all know and love. If you have heard of Atlas at all it is probably for the punishment meted out to him after the Titans were defeated by the Olympians: he was condemned to hold up the sky for all time. The image of the bearded, be-muscled Atlas with his head bent beneath the weight of the celestial sphere on his shoulders is probably one you can summon to mind from statuary and art both ancient and modern. However, Atlas also features in another more mountainous ancient story. As told by Ovid, Atlas was a king with an orchard of golden apples, which he had been told would one day be stolen by a son of Zeus. One day, the hero and demigod Perseus arrived and asked him for shelter. Suspicious, for Perseus was Zeus' progeny, Atlas refused, and so the hero – handily having the head of Medusa in his sidebag – turned him to stone. The king metamorphosed not into a human-sized statue, as one might expect, but into a mountain. (The golden apples Perseus left untouched; they were eventually stolen, generations later, by another son of Zeus, Heracles.)

The image of Atlas being turned into a mountain inspired several artists over the course of the sixteenth and seventeenth centuries, often by way of printed illustrations to editions of Ovid's *Metamorphoses*. An Italian engraver, Antonio Tempesta (1555–1630), places Atlas as a life-sized figure atop an existing pinnacle, with Perseus performing a Medusan fly-by on his winged horse Pegasus. As he does so, Atlas's skirts ossify into a sheer cliff-face, but his arms and expression of shock and surprise are still flesh and blood. In this illustration, the viewer is up in the heights with (or of) Atlas, and a wide-open landscape of smaller hills recedes into the distance, with just the suggestion of buildings to allude to the city of which Atlas is king.

Another depiction of the myth (fig. 14) dates to 1685. It is not straightforward to name the 'artist', since illustrations engraved in one decade were often re-engraved for later editions. This particular image started life as a design by a German engraver, Johann Wilhelm Baur (1607–1640), and was then re-engraved by a Frenchman, Abraham Aubry (died 1682). In this version, Atlas is huge, the same size as the mountain into which he is transforming. He is leaning into his mountain as if into an enveloping beanbag, his shoulders forming the upper slopes of the mountains and his head the peak. The contours of his human body can still be seen, but his left foot has already vanished, transformed into the stone roots of the hill. Next to him, Perseus is tiny, and holds up the head of Medusa almost casually. Atlas has one hand on his hip, the other holding a sceptre of some sort. This Atlas does not look surprised so much as offended that the puny Perseus should have the gall to entrap him in such a way. This 'Atlas mountain' (and there is a real-life mountain range named after this mythological character) has as much personality as Michael Drayton's Skiddaw.

Mountains Before Mountaineering

14. Atlas transformed into a mountain, Abraham Aubry after Johann Wilhelm Baur (1607–40). (Science History Images / Alamy Stock Photo)

The final category which emerged from my inventory of mountain artworks was that of religious images which incorporated mountainous scenery. Of course, the Holy Land is a place of mountains, many of which, as we know from Jean de Thévenot, formed the focus of seventeenth-century pilgrimage and tourism. It should not come as a surprise to find mountains in the background of artistic depictions of Biblical events. What is surprising, however, is the specific type of images which mountains most frequently featured in, or did not feature in.

Over the course of my search I found no early modern depictions of Mount Quarantine, or Jesus' temptation by the Devil. Paintings of Moses' retrieval of the Ten Commandments from the summit of Mount Sinai were few, for all that it appeared in miniature in the pages of the *Atlas Maior*. There are some images of the Crucifixion (which took place on the Hill of Calvary) with mountains in the background, and of the Transfiguration, which occurred on the side of a mountain, later identified as Mount Tabor (575m). I also found mountains in artworks depicting either the flight of the Holy Family to, or return from, Egypt, where they temporarily resided to avoid King Herod's slaughter of every male child under 2 years of age. One lovely oil painting, Karel Dujardin's *Return of the Holy Family from Egypt* (1662), is divided by the vertical rule of thirds taught to many a modern landscape photographer (fig. 15). In the bottom third, the Holy Family is crossing a river. Joseph is bending over a donkey whilst Mary is pointing to the sky and looking down, in an attitude of maternal exhortation, at Jesus, shown here as a small, curly-haired boy apparently dragging a sheep through the ford by its ears. The upper third of the painting is a blue sky with white clouds, and in the middle third a sheer, craggy mountain, some distance away, shadows the trio. Does the mountain stand for their exotic exile in Egypt, or the distance they have travelled or must travel before they can return home?

However, what struck me above all was the frequency with which mountains were incorporated into depictions of the Virgin and Child. This is one of the most significant images in Christian iconography, capturing both God's love in sending his son to Earth to save mankind and Mary's tender maternal care of her divine offspring. As backdrop to this most comforting of scenes, mountains appear again and again. Sometimes, if the pair are depicted inside a room, a single peak can be seen through a window. Where the pair are shown outside, a peak might rise as a subtle, distant feature, over the top of one of Mary's shoulders. In other depictions, mountains make up a more significant proportion of the paintings. One oil-on-copper from 1600, by an anonymous German artist working in the style of Albrecht Dürer, shows the holy pair in the foreground of an inhabited Alpine landscape. Buildings skirt the shore of a lake, with nearby wooded crags giving way to more distant, rocky ridges in the background.

The image which struck me the most, however, was Marco d'Oggiono's *Virgin and Child Enthroned with Saints* (fig. 16), a large (176 x 148cm) oil on canvas dating from around 1524. Mary is in the centre of a grouping of three saints; the baby Jesus is leaning out from her lap to either bless or grab at the beard of one of them. Mary is seated upon a pedestal, and behind her rises an almost triangular mountain peak. The effect is such that she seems to be enthroned not just by the chair upon which she sits but by the mountain itself. Behind this closer peak a blue mountain range bisects the upper third of the painting. The striking nature of this artwork depends entirely upon the inclusion of mountains: take them out and for all the charm of the human scene depicted, it would not draw the eye in the same way at all.

15. Karel Dujardin, *Return of the Holy Family from Egypt* (1662), oil on canvas, 62.6 x 51.1cm (c. 1662). (Photo gift of James E. Scripps © Detroit Institute of Arts, USA/Bridgeman Images)

16. Marco d'Oggiono, *Virgin and Child Enthroned with Saints* (c. 1524). (Vidimages / Alamy Stock Photo)

At the most basic level, then, mountains helped to provide visual framing. But as I found example after example like this, I wondered whether their inclusion in paintings of the Virgin and Child meant something more. Were they simply a reflection of the fact that both the Holy Land and many of the artists' home landscapes were hilly or mountainous, or did they symbolise something more? I am inclined to conclude that it was a bit of both. Depictions of the Virgin and Child were popular and treasured because they represented the *start* of the most important story in Christianity: Jesus' incarnation as a human that he might grow up and one day die on the Cross. An important later episode in his story was the time he would spend in the wilderness, and his temptation by the Devil on Mount Quarantine. My theory is that in many of these paintings the mountains are a shorthand for that wilderness. However, wilderness was explicitly not a bad thing within the life story of Jesus: instead, just as in the story of Theuerdank, it was formative. The mountains in the backdrop of images of the infant Jesus might therefore allude to the potential of the mountain space to help transform the child into the man who would, in the Christian worldview, save the souls of all mankind. Or, these 'heaven-shouldering' peaks, as Joshua Poole would term them, might have been intended by the artist to put the viewer immediately in mind of the heavenly God as well as his earthly incarnation.

Whatever the answer, one thing is obvious. If you were a sixteenth- or seventeenth-century artist, painting the most beloved image in the Christian canon, you would not incorporate a landscape feature that would cause your viewer to recoil with disgust or fear. Whether they served a specific symbolic purpose or not, mountains added to rather than detracted from the beauty and appeal of these images of maternal love and divine grace. More than anything else, these paintings of the Virgin and Child confirm for me the basic idea that they *did* like mountains back then.

So, the meanings of mountains as found in early modern art and literature are both similar and different to the meanings of mountains today. Many of the 'tropes' found in the *English Parnassus* would still ring true today: mountains are high, cold places, and the sunrise across them is a sight to be seen. Theuerdank and Brennus show us that mountains, just as today, were viewed as sites of physical endeavour, heroism and danger, but with a specifically early modern twist: mountains helped test princes, rather than elite climbers. Mountains also had meanings which they have largely lost today. The most famous mountains were not the highest ones, but rather those most closely attached to ancient myths and legends. A well-read traveller venturing along a mountain path might have reasonably thought of nymphs and muses, not in any real expectation of meeting one, but in much the same way my own thoughts would turn to George Mallory and Sandy Irvine if I found myself looking up at the North Face of Everest. Mountains even sometimes spoke for themselves, and when they did they stated quite clearly what they expected: to be a place for your delight.

Vantage Point: Into the Volcano

What kind of a person does it take to stand at the edge of the crater of an active volcano, have a coughing fit from the smoke rising from the lava, and feel a sense of delight? It turns out the answer is a German Jesuit monk who was fascinated by everything from the instrumental properties of felines to the inner workings of the globe.

Called 'the last man who knew everything' by his modern-day biographer, Athanasius Kircher (1602–1680) was not only a man of wide interests but also deeply learned.[1] His work on Egyptian hieroglyphics, though it proposed translations which have been shown to be total fictions, established the link between ancient Egyptian and Coptic which would prove to be essential in the eventual decoding of the Rosetta Stone. His *Arca Noë* (i.e. Noah's Ark) theorised on the exact required dimensions of the ark, the necessity for carrying extra livestock to feed the lions, and even how the ark was likely to have been loaded (humans and birds on the top deck; four-legged animals including unicorns on the bottom). He was also an inventor and fascinated by strange technologies. In 1650, he famously described a 'cat organ' whereby pins, attached to keys, would strike the tails of a series of boxed cats whose 'natural voices' had a variety of pitches and whose yowls of pain could thus be transformed into a melody. This instrument, Kircher reported, was first devised to lift the spirits of a downcast Italian prince (clearly not a cat-lover).

The cat organ, or *katzenklavier*, appeared in his *Musurgia Universalis*, an encyclopedia of music spanning over 1,100 pages.

Given all this, it would be more of a surprise to find that Kircher had not written about mountains at some point than that he had. In his *Mundus Subterraneus* (1665) he set out to explain nothing more or less than how the world worked. It included the first printed charts of oceanic currents, Kircher's theory on the location of the island of Atlantis and detailed explanations of the hollow spaces beneath the surface of the Earth. According to Kircher, the world was filled with 'Treasuries of Fire', which were given the opportunity to escape to the Earth's surface through the mechanism of volcanoes. In the early modern period, volcanoes were viewed as a sub-category of mountains. They were often described as 'fiery mountains' or 'burning mountains'. It was theorised that both mountains and volcanoes were hollow, with the only difference between them being whether their insides were filled with water or fire.

It is not clear whether Athanasius Kircher ever saw a cat organ in the fur, but it is certain that his writings about volcanoes were informed by real-life experience. In 1638, he climbed up and into the craters of the two most famous volcanoes in Europe: Etna and Vesuvius. Kircher believed that the stores of fire in the centre of the Earth and the volcanoes which released them on to the Earth's surface had a very important purpose: they would provide the means by which the world would one day be destroyed as promised in the Bible. The surprising thing is that when climbing these portents of ultimate worldly destruction, Kircher found the idea of the fiery end times inspiring rather than horrifying. Peering into the depths of Mount Etna, he was struck by the noise of the volcano's 'roarings and bellowings', which were so loud 'that they make the very mountain it self to quake and tremble'. However, his very next comment was one not of fear but of almost religious ecstasy:

> In a word whoever desires to behold the power of the only great and good God, let him betake himself to these kind of mountains; and he will be so astonish'd and stupefied with the ineffable effects of the miracles of nature, that he will be constrained ever and anon to pronounce ... *O the depth of the riches and wisdom of God!*[2]

Kircher was quoting from the Bible (Romans 11.33) and he was awestruck by the thought that volcanoes represented even a tithe of the greatness and grandeur of the God to whom he had dedicated his life. On his return from Mount Etna and Sicily, a series of earthquakes caused the sea to become so rough that his ship was forced to make land unexpectedly on the shores of Italy. Kircher did not mind the life-threatening chaos, since it gave him the opportunity to visit yet another volcano, 'the famous *Vesuvius*'.

At the foot of the hill, Kircher hired 'an honest countryman, for a true and skilful companion, and guide', though he had to offer an 'ample reward' to this mountaineer to lead him at midnight up 'difficult, rough, uneven, and steep passages'. Reaching the crater, Kircher's senses were overwhelmed by a scene which resembled nothing so much as 'the habitation of hell'. The sight of fire, the stench of sulphur, the clouds of smoke which the mountain 'belched' out towards him and which almost made him 'vomit back' at it. What was Kircher's response to this apparently ghastly landscape? Once again, it was wonder and awe at the capabilities of God. If the creator of the Universe gave even an inkling of his power in the great fires of Vesuvius then what, he wondered, would it be like 'in that last day, wherein the Earth shall be drowned in the ire of thy fury, and the elements melt with fervent heat'?[3] Again, Kircher was not distressed by the thought that volcanoes offered a taste of the end times; instead, he was giddy with it.[4]

Athanasius Kircher was an extraordinary individual, which is another way of saying that he was a genius who was also undoubtedly somewhat mad. However, in his response to volcanoes he was actually in keeping with the educated elite of his time. He saw the world as an opportunity to better understand the nature and mind of God Himself, and viewed the end of the world as the culmination of the whole cycle of human history, which had started with Creation and the fall of Adam and Eve and would end with the destruction of the original Earth and everlasting redemption on a new Earth. Volcanoes – fiery mountains – were an object of delight and amazement because they offered a foretaste of much-desired things to come.

Chapter Four

Mysteries of Science, Mysteries of Faith

If you are reading this book then you have probably, at some point or another, looked up at a mountain and enjoyed the sight of it; you might have thought it beautiful, or even felt thrilled and inspired by the height of it towering above you. I do not expect that your moment of aesthetic enjoyment would have involved much deep thought on where mountains came from (unless you are a geologist) or whether you were being theologically correct in admiring them in such a way. In this chapter we will explore the ways in which science, religion and the beauty of mountains overlapped in surprising ways during the seventeenth century.

Religious science, scientific religion

The term 'science' is a slightly awkward one to apply to the early modern period. The discipline, as we understand it, had yet to take its modern form. Imagine you are at a party – perhaps a bit like my dad's retirement party where I got quizzed about mountains and when people started to like them – and you enter into conversation with a seventeenth-century scholar who has accidentally fallen through time. He tells you that he writes books about how light, water and gravity work, or even about how the universe came into being. 'Oh!' you say, 'so you're a scientist?' He'd probably look at you blankly. He might say no, he's actually a schoolmaster, or a vicar. You press him: but you study science? Oh no, he'd say, he studies natural philosophy. Or he might even say that he studies 'the works of Nature and God' (and you'd be able to hear the capitalisation of 'nature').

Later, over the snack table, you might encounter this individual again, and ask him exactly what it means to study natural philosophy. Whilst he cautiously takes his first nibble of the modern novelty of Pringles, he'll tell you about walking the hills in search of unusual plants and animals but also about reading books of ancient Greek and Latin and above all poring over the Bible in order to understand how the world works. He might tell you about alchemical experiments, and the various glass containers used to distil different materials, and you might be put in mind of modern chemistry. But the early modern natural philosopher did not confine himself solely to experimentation and observation. History and theology were just as important to the theories he developed about the world. The division between different 'disciplines' or subjects of study, according to which modern-day education is organised, did not begin to crystallise until the eighteenth century. Instead, most scholars were, like Athanasius Kircher, jacks of many trades.

Another division which did not exist in the early modern period was that between science and religion. Instead, religion underpinned all science (or natural philosophy) because belief in God permeated almost all aspects of the early modern life experience.

How important is religion today? It is a thorny question, and bald statistics can only take us so far in answering it. In the 2011 UK census, just under 60 per cent of respondents identified themselves as Christian. The next largest slice of the pie chart, at just under 26 per cent, was 'no religion', followed by a 4.4 per cent slice for Islam and smaller slices – all less than 1 per cent – for Sikhism, Judaism and Buddhism. (And yes, a number in the hundreds of thousands also answered 'Jedi'.) Of course, 60 per cent identifying themselves with Christianity does not mean a population made up mostly of practising Christians: I suspect most churches would be giddy with excitement if weekly numbers represented over half of the members of their parish.

For many, religion intrudes only occasionally into their daily lives – carols at Christmas, maybe attending a service on Easter Sunday at a push. I myself would probably put 'Christian' on the census: I went to church as a child, sang in a chapel choir at university and enjoy a good carol service. In a cultural sense, I feel that Christianity is important to me. But in terms of actual beliefs I am foggily agnostic, and religion has relatively little impact on my day-to-day experience of the world. I mostly think about the Bible in reference to my historical research: all those years at Sunday School have paid off mostly in my ability to recognise Scriptural references in early modern writings.

In seventeenth-century Europe, the situation was quite different. Perhaps even later on in the party, with only a few guests left and stronger alcohol brought out, someone might ask '*Is there a God?*' and that time-displaced natural philosopher would crinkle his brows. Well, he might say to the room, there are certainly philosophical arguments which question the precise nature of God, and some which might even challenge the idea that He exists – but these are merely intellectual exercises. If someone then interrupted and asked him, yes, but do *you* believe in a God, then he would say, of course. Where would you even begin understanding the universe if you didn't?

And herein lies the most important and distinctive aspect of early modern efforts to trace the history of the natural world: they started and ended with God. Science was not separate from religion because science *was* religion, and religion was science. The reason natural philosophers wanted to understand the world was because it had, in their innate worldview, been made by God. Scholars thought about the world as 'the Book of Nature'. This corresponded to the literal book in which God had inscribed, in words, His revelations for mankind: the Bible, or what they would have called Scripture. In the Book of Nature, He had left more revelations for scientists, armed with the rational thought with which He had also gifted them, to uncover and interpret. Or, as one late seventeenth-century *Natural History* put it, 'every flower of the field, every fibre of a plant, every particle of an insect, carries with it the impress of its Maker and can … read us lectures of ethics or divinity.'[1]

The problem with the Book of Nature was that it needed to be interpreted, and just because natural philosophers agreed on the basic point that God had created the Earth did not mean they agreed on the details of *how* He had done so. Indeed, their deeply held religious beliefs meant that the controversies which erupted around these details could often be bitter and fierce. This is not a novel observation on my part. June Goodfield and Stephen Toulmin summarised it perfectly in their 1965 monograph *The Discovery of Time*: the debates of early modernity were 'not between science and religion, they were *within science*, as men then conceived it' – the ultimate goal of that science being to better understand God.² The term 'atheist' was one of the ultimate insults that one could throw at a fellow natural philosopher for going too far in their attempt to explain the workings of the divine.

Religious certainties, however, only went so far. The seventeenth century is a particularly fascinating period to study because natural philosophers who were interested in the origins of the world found themselves being forced to question many of the basic assumptions upon which their predecessors of the fifteenth and sixteenth centuries had happily relied. At the beginning of the Renaissance, in the fifteenth century, it was widely thought that the world was approximately 6,000 years old – a figure reached by charting the lifespans of the Old Testament patriarchs (Adam, his sons, Methuselah, Abraham, etc.). It was also generally agreed that the Bible literally described the process of Creation and that the world had undergone relatively few physical changes between the seven days in which God called it into being and the current time.

During the seventeenth century, however, new insights and ideas came to light which destabilised these comfortable assumptions. Investigations into fossils, alongside Nicolaus Steno's observation of stratigraphic layers to the Earth's surface, cast doubts on the literal truth of the Biblical account of Creation. Did stratigraphy indicate stages of Creation not described in Genesis? Had God made mistakes which He had been forced to discard creatures in fossil-form and then correct? Meanwhile, increasing familiarity with civilisations beyond Europe threatened the hitherto-accepted age of the Earth. How to fit the recorded history of China, spanning far more than 6,000 years, within the traditional chronology? Questions tumbled upon questions: did 6,000 years even leave enough time for the descendants of a single couple – Adam and Eve – to populate the entire world?

It is easy, as a twenty-first century reader, to shrug and see these as fairly trivial details. After all, we are very accustomed, today, to thinking of the age of the Earth in terms of billions of years; to accepting that the portion of time covered by written human history is but the narrowest and most recent slice of the story. However, the gulf between the current-day mindset and the early modern one is huge. Paolo Rossi (1923–2012), an Italian historian of science, wrote of the scale of this gulf. On the one hand stood the comforting sense of knowing, from the Bible, the entire history of the world, and on the other stretched the 'dark abyss' of deep time, its history mostly unrecorded. On the one hand stood the belief that you lived in a world which matched that 'shaped by the benevolent hands of God', and on the other the stomach-churning suspicion that the world as you knew it bore no resemblance to the original Creation.[3] This in turn forced scholars to question the unquestionable: how true was the Bible? If it could no longer be seen as literally true in all aspects, how then should it be interpreted?

Thus, the scholars in this chapter, when they tried to truly understand and explain mountains, where they came from and what they meant – whether they should inspire disgust or awe – found themselves in a precarious position. They stood on the edge of the 'dark abyss of time' and were forced to map territory uncharted by previous generations, to try to bridge the gap between what they believed about God and what they thought about the natural history of the world. Given all this, it is not surprising things got as heated as they did.

Thomas Burnet's contradictory history of mountains

In the introduction, I talked about my professor who called the seventeenth-century natural philosopher Thomas Burnet a 'friend' of his. I suppose by this stage I should claim him as a friend of my own, since he has now been part of my life for a decade and counting. How to introduce him, as if he were that time-travelling party guest and I the only person in attendance who knew him? This is Thomas, he runs a school and has written a book about the history of the Earth. Later, if he got heated in some discussion or other, I might shrug in apology: yes, that's Tom, he really does stick to his guns when it's something he cares about.

In conversation with Burnet you might learn some facts about his life. He might boast of how young he was – a mere 22 years old – when he became a fellow of Christ's College, Cambridge, in 1657. He might share anecdotes of the time when, fresh from his own studies, he served as tutor to a young nobleman as he travelled across the Continent. He might tell you that he visited the Alps and about the turmoil the sight of mountains threw his thoughts into.

Since he is not here in person, I would like to share a few more biographical details on Burnet's behalf. He was elected master of Charterhouse School in 1685, a role he would hold until his death. (Charterhouse exists as an independent school to this day and, in a strange historical coincidence, provided employment as a teacher to George Mallory before and in between his Everest attempts.) Not long after his election, he fell into a dispute with the king. James II, a Catholic, wanted students to be able to enter the school without swearing oaths to the Church of England. Burnet, his Protestant principles clearly stronger than his sense of royalism, vetoed the idea. A year later, King James II was ousted and replaced with his Protestant daughter Queen Mary and her husband William of Orange, and Burnet rose in popularity at court. He was even mooted as the next Archbishop of Canterbury, before falling out of favour again. He never married and died at Charterhouse at the ripe old age of 80 in 1715. He is most famous today for his *Sacred Theory of the Earth*.

This is an account of Burnet's life, but what was he *like*? Whilst in Rome in 1675, he was painted by the esteemed portrait artist Jacob Ferdinand Voet. He looks, to be honest, a bit of a poser in this painting: dressed in an enviably lavish silk gown, with long brown curls (probably a wig), his left arm is casually draped across his chest and his thoughtful expression would rival that of any desperate romantic. Some years after the portrait was painted, another hand added lettering around its oval top, identifying Burnet as the author of the *Sacred Theory*. Around twenty years later, Burnet would be painted again, this time by the German-turned-British court painter Sir Godfrey Kneller (originally Gottfried Kniller). This is a far more sombre Burnet – at work rather than on holiday (fig. 17). He is in plain schoolmaster's robes with white clerical bands around his neck, and his hair is his own, falling with a slight wave to just below his ears. Behind him, a book is open on a stand; it bears the distinctive frontispiece (opening illustration) of his *Sacred Theory*. He still looks thoughtful.

Being a historian is not just about finding material; it is about deciding how to write about the past, how to explain it to different audiences. Being honest, I have found it very difficult to write about Burnet and his *Theory* in this book. I know them both very well, so much so that I sometimes forget how complicated Burnet's ideas were. His *Theory* covers so many things, some of them contradictory. It tells about how he hated mountains, but also how they inspired him. It was built on his fear that Scripture did not make sense and represents his heartfelt attempts to solve that. His book confused people and it upset them.

17. Mezzotint portrait of Thomas Burnet by John Faber Jr (1752), reproducing an oil painting by Godfrey Kneller (1697). (© National Portrait Gallery, London)

Burnet's *Theory* was a book about the life cycle of the Earth: its past, present and future. It was first published in Latin in 1681 and in English in 1684, with second volumes in both languages published in 1689 and 1690 respectively. Burnet wrote it because he was worried. He was worried about the gap between what he, as a natural philosopher, could observe of the natural world and the implications of a literal reading of Scripture. He did not think nature disproved Scripture, but he worried that prevalent interpretations of key events in Biblical history – in particular Noah's Flood – simply did not make sense, and in so doing threatened the security of the entire Christian faith. He wanted to explain all the details of the Earth's development – from its original creation all the way through to its future destruction and the establishment of a New Earth – in a logical, rational way. He believed that in doing so he would help preserve religion from doubters and critics.

Burnet's first problem was with the Biblical account of the Flood, which he referred to as 'the Deluge'. He did not think the description in the Bible could be read literally. Unlike Kircher, he was not particularly concerned with calculating the size of the ark required to house a breeding pair of every species of living thing upon the Earth, or with the practical challenges of feeding and maintaining them. Instead, he was worried about the precise mechanics of a flood which covered the entire world. As he explained in his *Theory*, Burnet found himself stuck upon one point in particular: the fact that the Bible reported that the Flood covered the tops of the highest mountains.[4]

Why was this a problem? Because of the increasing uncertainties, mentioned earlier, facing early modern scientists. As they gained greater understanding of the world, more aspects of their Christian beliefs came under doubt. Even if they accepted that the Bible was not literally true in every detail, they still wanted to be able to prove that there were real-life explanations which fitted what was written in Scripture. If you could not do this, then maybe the Bible as a whole was not true, and that idea was not acceptable even to contemplate.

So that line in Genesis worried Burnet greatly because, as far as he could tell, all the water in the world would not cover the tops of the highest mountains. By his calculations you would need 'at least eight Oceans' worth of water to bring the water levels on Earth above the tops of the highest mountains. Earlier theories which had attempted to solve this problem did not satisfy him. One had suggested that the waters had emanated from 'the deeps' of the Earth but, as he fairly sensibly pointed out, there was nothing to stop all the water immediately draining back into the deeps. Another explanation suggested that God created more water for the Flood and then destroyed it afterwards. Burnet did not like this answer either. He felt that an explanation which saw God going 'forward and backwards' was an insult to His wisdom. As an omniscient God, He would have known from the beginning that the Deluge would one day have to occur.[5]

So Burnet had problems when he studied the Bible. He was also troubled and intrigued when he looked at the world around him – and particularly at mountains. It was his journey across Europe as a young tutor which partly inspired him to devise his *Theory*. He explained that:

> There is nothing doth more awaken our thoughts or excite our minds to enquire into the causes of such things, than the actual view of them; as I have had experience my self when it was my fortune to cross the Alps and Appennine mountains, for the sight of those wild, vast and indigested heaps of stones and earth, did so deeply strike my fancy, that I was not easy till I could give my self some tolerable account of how that confusion came [to be] in nature.[6]

Mountains did not strike Thomas Burnet, as they had the poet Thomas Churchyard, as 'noble, stately thing[s]'. Nor did he find himself, as Thomas Coryate had, delighted by the sensation of being above the clouds. He looked at them and saw confusion and disorder, which he could not believe had been designed and created by God. His *Theory* ultimately killed two troublesome birds – mountains, and the Scriptural account of the Flood – with one stone. Mountains, he declared, did not exist until after the Deluge. In fact, they were the result of the very same process by which the Flood occurred.

To understand that process we must follow Burnet through the whole life cycle which he charted for the Earth, from the moment of creation to the end of days. It started out, he declared, as a 'Mundane Egg'. This was a globe (slightly elongated at either end, just like an egg) made up of a series of layers. The 'yolk' was the 'central fire', which was then encompassed by a heavy layer of solid matter, upon which settled all the water in the universe. From the very outset, Burnet wanted to explain things rationally: the Earth was structured this way because it made sense for matter to organise itself from densest to lightest. The motion of all this matter, in something called the primordial 'Chaos', had thrown up small particles of soil, which, taking longer to descend than the other elements, ultimately formed a thin crust over the surface of the waters. The lightest element, air, took its natural place circulating around the outer crust. This formed the original, or 'antediluvian' (pre-Flood) Earth, which was 'smooth, regular and uniform, without Mountains, and without a Sea'.[7]

This level Earth was, Burnet insisted, a real paradise: there was 'not a wrinkle, scar or fracture in all its body' and the 'smoothness of the Earth' meant that the air was likewise 'calm and serene' with none of the tumultuous storms 'which the mountains and the winds cause in ours'.[8] It was also perfectly structured to enable the occurrence of the Flood without any further intervention from God. Instead, the sun beating down upon the surface of the Earth would warm the waters below the hollow shell of the 'egg'. Eventually, at the 'time appointed by Divine Providence', the waters would begin to boil. This boiling would create not just a great 'commotion and agitation' but would also see the water transform into gas which (as Burnet quite correctly pointed out) has a greater volume than a liquid or a solid. The resulting pressure would at long last shatter the eggshell. Some parts of the Earth's crust would fall beneath the surface of the waters, and others would be thrown upwards. The waters would be in a tumult, with waves and surges that would truly 'cover' the tops of the newly born mountains.[9] When the waters subsided, they would reveal the world as we know it today, made from the 'ruins' of the first.[10]

Burnet did not think much of this second iteration of the Earth. With its ragged coastlines, irregularly sized seas and oceans and its mess of mountains, it represented nothing less than 'the true aspect of a world lying in its [own] rubbish'.[11] Fortunately, it would not last. In the second volume of his *Theory*, Burnet turned from the first destruction of the Earth by water to its final destruction by fire: the 'Conflagration' which would presage the Second

Coming of Christ, the Last Judgement, and the ultimate reward of the 'new heaven and the new earth' (Revelation 21:1) for those judged worthy.

Just as with the Flood, Burnet believed that God had prepared nature from the beginning for this second great disaster. Mountains again had a starring role. Every volcano in the world would erupt simultaneously, and 'new mountains in every region', hitherto quiet, would 'break out into smoke and flame'.[12] This global eruption would see the great cities and the great mountains of the world melting away: Rome and the Alps alike would be nothing more than wax in the face of such terrible heat.[13]

Finally, just as the Flood had created the Earth in the form that we know it today, the Conflagration would result in the transformation of the Earth into the shape it would hold for the rest of eternity: that of the new Earth, the final paradise for the blessed. What, you might ask Burnet, would this new Earth look like? Well, it would be 'of an even, entire, uniform and regular surface, without mountains or sea'.[14]

※ ※

Through his *Theory*, Burnet hoped to protect the truth of Scripture and preserve the foundations of the Christian faith. There was one casualty: the simple enjoyment or admiration of mountains. His whole theory was based on the presumption that mountains were disordered and ugly, and its ultimate implication was that mountains were a lasting reminder of the Flood, a disaster visited upon mankind for their overwhelming sinfulness. If as a reader you accepted Burnet's account of the life cycle of the Earth, then you would have little choice but to look at a mountain and feel disgusted.

Burnet dedicated his eleventh chapter to 'the Mountains of the Earth, their greatness and irregular Form'. In it, he acknowledged that his theory took something away from the experience of mountains. He opened with an extraordinary musing:

> The greatest objects of nature are, methinks, the most pleasing to behold ... there is nothing that I look upon with more pleasure than the wide sea and the mountains. There is something august and stately in the air of these things that inspires the mind with great thoughts and passions; we do naturally upon such occasions think of God and his greatness, and whatsoever

hath but the shadow and appearance of INFINITE, as all things have that are too big for our comprehension, they fill and over-bear the mind with their excess, and cast it into a pleasing kind of stupor and admiration.[15]

However, his point was not that this was the correct way to view mountains. Instead, he was saying that mountains were so overwhelming that most people did not stop to think where they came from, or whether they were really worthy of observation. There is certainly a hint of arrogance in what followed: his implication was that whilst 'the generality of people have not sense and curiosity enough to raise a question concerning these things', Thomas Burnet possessed both qualities in spades. It was his self-appointed job to reveal the truth about mountains to those less fortunate. The admiration of mountains was rooted in 'ignorance', but a greater pleasure was to be found in understanding their real origins and nature. His account of mountains did more, he insisted, than help to explain the Deluge: it was also the only explanation for the origin of mountains that made any sense.

He concluded the chapter with an illustration: a map of the world without borders or cities in order to better view the topographical features of the Earth, its mountains and coastlines (fig. 18). It is perhaps not the most striking sight to a modern eye, accustomed as we are to maps of many different types. Try to see it, however, through Burnet's eyes: as an image which highlights all the irregularities in the Earth's surface. He dreams of his map made three-dimensional, into a globe. Unlike the smooth globes in widespread use by the seventeenth century, his would be rough and textured, pitted and pockmarked. This 'being done with care and due art' would be a 'true model of our Earth'. It would reveal to the viewer the truth: 'what a rude lump our world is which we are so apt to dote on'.

AN EARLY MODERN FLAME WAR

Burnet and his *Theory* got under a lot of people's skin. To get a sense of the emotionally charged debate which he sparked it is worth looking in depth at his lengthy interchange with one Erasmus Warren (*c.*1656–1718), rector of the village of Worlington, Suffolk. His first critique of Burnet, published in 1690, was a fat book entitled *Geologia*. Burnet swiftly produced, in the same year, a slim volume responding to Warren's criticisms. In 1691, Warren published a 'full reply' to Burnet's response, to which Burnet again penned a hasty rejoinder. Finally, in 1692, Warren published his last words on the matter: *Some Reflections upon the Short Consideration of the Defence of the Exceptions Against the Theory of the Earth*. I would say you have to feel pretty strongly about something to find yourself writing and publishing a third-order response to it.

This increasingly ludicrous interchange represented the heart of the 'pamphlet war' surrounding Thomas Burnet's ideas. Such published wars of words were a feature of the sixteenth and seventeenth centuries, when the increasing

18. The 'rude lump' of the world, with mountains emphasised, from Burnet's *Theory of the Earth*, facing p. 98. (Reproduced with kind permission from the University of St Andrews Special Collections)

use of the printing press enabled scholars with opposing views to air their differences in a public, permanent forum – much as the modern-day internet has done on another scale entirely.

What was Warren's problem with the *Theory of the Earth*? In general terms he believed that Burnet, rather than protecting religion as he had hoped, had instead assaulted 'the very foundation of it'.[16] Warren also specifically and emphatically objected to Burnet's description of the original Earth as being without seas or mountains – and his implication that the absence of mountains was the superior state of affairs. He compared the 'smoothness' of Burnet's proposed primeval Earth with the ornamentation of the current world, with its 'raised work, of hills; the embossings, of mountains; the enamellings, of lesser seas; the open-work, of vast oceans; and the fret-work, of rocks'.[17] Warren did not agree that a regular landscape was more attractive than an irregular one, arguing that 'the beauty of the Earth' lay precisely in its variety and contrasts, in the fact that in some places it contained seas and lakes and in others 'hills and mountainous roughnesses'.[18]

Burnet was quietly scathing about people who never looked beyond their enjoyment of mountains to question their origins. Warren turned this on its head, implying that people who could not look beyond the irregularities of mountains to see their beauty were fools. 'Truly,' he commented 'that roughness, brokeness, and multiform confusion in the surface of the Earth' may well seem to the 'inadvertent' (by which he means uninformed or misguided people) to be 'nothing but inelegancies or frightful disfigurements'. However, 'to thinking men', they more rightly appear as the 'carvings, and elemental sculptures; that make up the lineaments and features of Nature, not to say her braveries'. In this passage, Warren is communicating two things: that he thinks Burnet is an idiot; and that people who are not idiots quite rightly view mountains with approval and enjoyment.

Burnet's sense that mountains were disturbingly disordered led him to conclude that they could not have been created by God; by contrast, Warren elaborated that their beauty demonstrated 'the marvellous ... skill of her Maker, most rarely expressed'. That they were intentionally designed – rather than a mere by-product of a global disaster – was also evident to Warren thanks to their *usefulness*. Mountains, he pointed out, were useful insofar as they produced natural boundaries for nations, thus reducing conflict between different kingdoms; in producing rivers; as the source of different minerals; and as the habitat of 'innumerable wild creatures'.[19] This theme – of the usefulness of mountains – was one that responses to Burnet would pick up on again and again.

❧ ☙

As a teenager, I once spent two weeks shadowing a crown court barrister. I remember being struck by the way that in court barristers would presage a take-down of their opposite number's argument with the smooth phrase 'my learned friend'. In a nutshell, they would generally be saying something which equated to 'my learned friend is completely and utterly wrong' – akin to starting a highly disrespectful sentence with the phrase 'with all due respect'. Afterwards, in the robing rooms, these same courtroom opponents would revert to being genuinely cordial friends once the wigs were off, making jokes at each other's expense and heading straight for coffee.

The debate between Burnet and Warren was couched within similar dubious courtesies – but I have a feeling the pair would have been rather less likely to happily share a hot drink afterwards than the barristers. Warren opened his *Geologia* by referring to Burnet as an 'ingenious Author' and suggesting with qualified courtesy that as an abstract philosophical exercise the *Theory* was actually quite interesting.[20] Similarly, Burnet opened his first response to Warren with the statement that, 'if it be a civility to return a speedy answer to a demand or message' he would not 'fail to pay that respect' to Warren.[21]

By the time they both reached the second volumes of their argument, however, the gloves had come off. Warren excoriated Burnet as 'indiscreet, rude, injudicious, uncharitable' and ruled by the 'brats of passion' rather than by rational argument.[22] In return, Burnet accused Warren of bulking out his response with 'popular enlargements, juvenile excursions, stories and strains of Country-Rhetorick'.[23] Later, Burnet went for what, in the context of seventeenth-century scholarship, was a real sucker punch: he criticised Warren's Latin, implying that the reason he wrote his critique of the *Theory* in English was because he was incapable of writing in the ancient and more scholarly language.[24] (This is somewhat hypocritical, given that Burnet happily published an English version of the *Theory*.) Warren closed his third and final response to Burnet with a final low blow. He observed that Burnet had recently commented that another one of his scholarly opponents 'writ neither like a gentleman, nor like a Christian, nor like a scholar'. Warren wondered whether the man he had once termed an 'ingenious

Author' might consider applying those words to himself.[25] 'Burn', as the lingo of my own generation might have it.

The first time I read through this back-and-forth debate, I was astonished at the bitterness of Burnet and Warren's approach to one another. The opening pages of Burnet's first response give a hint that there is more going on here than a detached intellectual debate. In excusing the haste with which he returned his response to Warren, and thus any deficiencies in his style, he proclaimed that, 'such personal altercations as these, are but *res periturae* [passing things], which do not deserve much time or study.'[26]

So what was personal between Thomas Burnet and Erasmus Warren? I went digging and found something interesting. Unlike Burnet, who as both master of Charterhouse and, briefly, a darling of the court, left behind portraits and other clear traces of his biography on the historical record, Warren left relatively little mark beyond his published writings, which consisted of his tussle with Burnet and a handful of religious tracts. At last, in the matriculation registers of the University of Cambridge, I discovered an intriguing detail: Erasmus Warren was admitted as a sizar at Christ's College, Cambridge, in 1656, aged 14. Burnet, you may recall, became a Fellow of that same college in 1657. It is worth noting that a 'sizar' was the word for a poor student who, in return for being supported by the college, undertook menial tasks such as waiting on tables.

Upon discovering this I had to wonder: did Burnet – as a Fellow – ever teach Warren Latin, adding a deliciously catty note to his comment about the younger man's grasp of the language? Did the teenaged Warren, clearing plates from the High Table at which Burnet sat, suffer some disparaging comment from the new Fellow, puffed up with his own prestigious position? Did the memory of serving the arrogant young man I imagine Burnet to have been rankle with Warren for decades until he chanced to read the *Theory of the Earth*, upon which years of resentment finally boiled over into an ostensibly intellectual pamphlet war?

This is sheer speculation, at least in the details. I find it compelling because it brings into sharp relief how deeply emotional apparently 'scientific' debates could be. The mutual antipathy between Burnet and Warren may represent an extreme example of this. However, Warren was far from being the only seventeenth-century scholar to pen a deeply impassioned defence of the mountains which Burnet's *Theory* had tried to tear down.

The author whom Burnet had accused of writing 'neither like a gentleman, nor like a Christian, nor like a scholar' was none other than the bishop of Hereford, Herbert Croft (1603–1691). Croft was the first person to publish a response to Burnet, and of all of Burnet's critics he is probably the one of whom I am most fond. He was 81 when the English version of the *Theory* was released, 82 when he published his own take-down of it. Throughout his *Animadversions Upon a Book Intituled the Theory of the Earth* (1685) Croft gives off the unmistakable impression of a man who feels he is far too old for all this nonsense. In his preface, he scowled at 'the learned men of the universities', and their failure to take up their pens to 'confute the fables' published by Burnet.[27] Their oversight meant that 'tho now in the eighty-second year of my age', and by his own account half-blind, Croft was forced to rouse himself to write a 'short essay' in criticism of Burnet, in the hopes that it might inspire a younger person to take up the same cause.[28]

Croft approached Burnet's *Theory*, then, with the impatience and querulousness of an old man dragged reluctantly out of a peaceful retirement. He termed Burnet's ideas 'vain fopperies' and accused him of making the Bible into a 'nose of wax', to be moulded to his own theories. In doing so, Croft fumed, he had done nothing more than delight atheists, by 'making the word of God, whereon the truth of our salvation depends, so uncertain and questionable'.[29] A later participant in the debate, one Archibald Lovell, would go even further and accuse Burnet of being 'an infidel' in his mistreatment of Scripture.[30]

When it came to considering mountains, Croft was just as excoriating. He commented, sarcastically, that Burnet's perfectly smooth world:

> wants only one thing; there is not a mountain in all his world to carry you (as the devil did our Saviour) from whence you might have a large prospect of this delicious land: for then doubtless you would fall down and worship him for his admirable contrivance of it.[31]

Not only did this passage blatantly poke fun at the flatness of Burnet's original Earth: it also lampooned his arrogance, suggesting that his real desire was to see his readers 'fall down and worship him', just as the Devil wanted Jesus to do when tempting him on Mount Quarantine.

Croft immediately put his hand on the subtext beneath Burnet's aesthetic attitudes towards mountains. 'Our Philosopher,' as he caustically referred to Burnet, 'was so much unsatisfied with the misshapen appearance of our present Earth' that, rather than viewing it as 'the complete workmanship of a master-builder' he instead deemed it to be 'a confused heap of rubbish'.[32] Croft, however, was determined to prove that the current form of the Earth was the handiwork of God.

His first point was that there was nothing wrong with irregularity in shape and form. He demonstrated this with reference to the human body, in a deliberate parody of Burnet's critical description of the form of the Earth. For one on the lookout for deformity, Croft suggested, they might term the human head 'to be like a jug or bottle with the neck turned downwards', the face containing a 'hollow trunk of a nose', the hands 'ragged and jagged with fingers', and the interior of the body 'all hollowed with caverns'. Except nobody would question that the human body had been designed by God, and 'although the parts seem thus uneven and disproportioned, we greatly commend the beauty of them altogether'.[33] Likewise, 'this great body of the Earth taken all together hath a wonderful beauty and admirable structure, even in those parts which he sets forth as most disagreeing and deformed'.

The elderly bishop honed in on Burnet's statement that the sight of mountains inspired him and made him think of God. Croft agreed, and insisted that:

> surely all men who behold these things [mountains] have the same delightful contemplation, as he acknowledges to have felt, when he beheld them; and yet we have never looked upon them as broken ruined fractions of a former Structure, which we poor Souls never dream'd of, till his Theory gave us notice of them.[34]

Here he was effectively turning Burnet's thought process on its head. Burnet was trying to say that most people were mistaken in feeling such things about mountains. Croft's sarcastic implication was that it was much more likely that Burnet was the one who was mistaken.

Ultimately, Croft rejected Burnet's suggestion that the Earth in its current form – complete with its hideous mountains – was anything other than the original Earth, created by the hand of God, or that a smooth, sea-less and hill-less Earth would be in any way superior. Were he to be stuck on Burnet's original Earth, Croft concluded, he would desire nothing more than 'to be transplanted into this misshapen irregular world ... where we have the variety of delightful motions and prospects both by sea and land'.[35]

※ ※

The elderly bishop of Hereford was not the only reader of Burnet to defend mountains using comparisons to the human body. Indeed, some would argue that drawing analogies between different things was an essential part of the way early modern Europeans thought about the world. In his *Order of Things* (1966), the French philosopher Michel Foucault suggested that each period possessed its own 'episteme': essentially how people living in that period thought about truth. The modern period, for example, is an episteme characterised by the understanding that new knowledge is established by empirical scientific inquiry.

Foucault envisaged the 'Renaissance episteme' as rooted in what he called 'relations of resemblance': in essence, that truth about one aspect of the world could be divined through thinking about how it related to other things. One such 'relation' was *analogia*; the sense that it *meant* something for different parts of the natural world to be similar or analogous to each other.[36] Now, Foucault's vision of the Renaissance episteme has been roundly criticised, but I think he was on to something with his identification of *analogia* as being crucial to early modern natural philosophical thinking.[37] It certainly influenced how some writers in the Burnet debate thought and felt about mountains, and in ways which seem quite foreign to the modern 'episteme' in which analogy is a literary device rather than a route to scientific understanding.

As already hinted at by Croft, above, one way of thinking about the world in the early modern period was to envisage it as a giant body, for example with veins and arteries in the form of streams and rivers. Where, then, did mountains belong in the analogy of the Earth as body? One participant in the Burnet debate, John Beaumont, pointed out that 'the Ancients call'd the Earth ... our Mother Earth', and even 'compared the Mountains on the Earth, to the breasts of a Woman'. Beaumont thinks this is fitting on multiple levels: breasts are both beautiful and useful, since they not only succeed in 'beautifying' a woman but also in 'yielding sweet streams of Milk for the nourishment of her Children'. Mountains, likewise, are 'ornamental', and provide 'continual streams of fresh Waters' from their tops.[38] Remember the poetic examples in

Chapter 3 which compared breasts to mountains? This was not just a reflection of the fact that mountains and breasts (depending on the contours of both, I suppose) sometimes resemble each other. Or rather, it was, but that resemblance meant something more within the early modern mindset of analogy. If mountains look like breasts, and breasts are attractive, mountains must also be considered to be appealing, and we should also think about the ways in which they function (or were thought to function) like breasts in providing the world with life-giving liquid.

After the above paragraphs, it is perhaps hard to believe that it is now that we reach the most X-rated moment of this book, but here we are. Every now and then, when reading something written hundreds of years ago, your average historian comes across a passage that makes them choke on their coffee. I had one such moment when reading the words of one Matthew Mackaile, a Scottish doctor who has left behind no record of his life and personality other than a series of highly eccentric publications. Mackaile's response to Burnet was published in 1691, and was written as a 'prodromus', i.e. a forerunner to a longer publication, which in this case never came to fruition. It opened reasonably enough, including a prefatory poem which declared of mountains that:

> These usefull Swellings doe appear to Me,
> No Gastly, Monstrous, Ugly Things to be:
> And it is to the MAKERS Skill no Stain,
> To say, the Earth was ne're on spacious Plain.[39]

Mackaile objected to Burnet's *Theory* on four main heads. The first two were theological – he believed Burnet had overstepped the mark in seeking to explain everything about the mystery of Creation, and (ironically, since Burnet explicitly set out with the aim of avoiding this) that he had 'needlessly multiplied Miracles' in his account of the life cycle of the world. The second two, however, were focused on the question of the form of the Earth. Mackaile felt that Burnet had written 'most undervaluingly' of the Earth in its present form and, finally, that he was quite simply wrong in his claims that mountains were not part of God's creation. This final point, Mackaile states in his introduction, he would prove through 'Analogie with Mans body'.[40]

Now, one might reasonably assume that here Mackaile was using the word 'man' to mean 'human' (a vagary typical until barely two decades ago), but, as I discovered a few seconds before my laptop became covered with coffee, this was not the case. For, it turns out, Mackaile's main defence of mountains is as follows:

> He [Burnet] may alse [also] be displeased, at the Mountain-like Scrotum of a healthfull Man (which hath many Wirncles [wrinkles] upon it) in which are conserved, the Instruments, by which the Race of Adam, hath been preserved, or propogat upon the Earth.[41]

You can pause to let that sink in, along with the fact that Mackaile later felt the need to stretch his analogy yet further by terming the Peak of Tenerife (the highest mountain in the world, you will remember) the '*Penis Terræ*', or the penis of the Earth.[42]

Essentially, Mackaile is coming at the idea of analogy from the opposite direction to Beaumont, who highlighted the resemblance of mountains to objects generally believed to be beautiful, breasts, as proof of the beauty of the peaks of the Earth. Here, Mackaile was latching on to Burnet's suggestion that visible irregularity meant that an object was not fit to have been created by God. The underlying logic is that if an object in the natural world serves a purpose, it must have been intentionally designed. And 'aha' Mackaile was saying, quite literally striking Burnet beneath the belt: 'whatever the organ in question looks like, you can't deny that it's pretty damn useful.' And if the penis was undeniably designed by God, so too were mountains, because they resemble each other. Logical? Maybe not to the modern mind, but certainly to the early modern one.

Beautiful usefulness

The deeper I looked into the Burnet debate, the more I came to realise that 'it's pretty damn useful' was also a statement about beauty and aesthetic appeal. You may recall that Erasmus Warren, amidst his numerous criticisms of Burnet, talked about the usefulness of mountains. The resemblance of breasts to mountains, as highlighted by Beaumont, was a matter of function as well as form, and he identified other ways in which mountains were of 'use' to the world. For one thing, they offered a different type of environment from other landscapes, meaning that, thanks to mountains, the Earth could support a greater variety of both plant and animal species than Burnet's boring, smooth plain. Beaumont also argued (echoing Churchyard's *Worthiness of Wales*) that the mountain air helped to produce particularly admirable samples of humankind, making 'the Inhabitants of Mountains ... stronger of Limb, healthier of Body, quicker of Sense, longer of Life, stouter of Courage, and of Wit sharper than the Inhabitants of the Valley'.[43] Finally, mountains also (supposedly) protected the lowlands 'from the violence of blasting and fierce Winds', prevented enemies from invading, and offered higher land from which 'Hay, Corn, Cattel, Houses, and Men' could be preserved from the floods which brought destruction to the low plains around rivers.

What, however, has all this to do with aesthetics or beauty? When giving presentations or lectures about my work, I often have a slide showing a battered old pair of hiking boots which once saw me up many a mountain. I pair with this the observation that today, to call something 'utilitarian' is a nice way of saying that it is ugly, but that it doesn't matter because it does its job. So, from a modern perspective the repeated insistence of my early modern natural philosopher friends that mountains were useful seems a lot like damning them with faint praise – particularly when compared with the dizzyingly positive modern language identifying mountains as 'sublime'.

The problem is that early modern aesthetics were not the *same* as modern aesthetics. What humans think of as beautiful or attractive has changed throughout time. Traditionally, mountains have been held up as an example of this truism – with the supposed shift from premodern disgust to modern awe demonstrating how enormously aesthetic attitudes can change between centuries. Of course, my entire point in this book is that this was not the case: I believe that mountains were objects of admiration throughout the early modern period (and long before). But the aesthetic principles according to which people admired them have definitely changed.

Early on in my research, I spent a lot of time reading about aesthetics, and particularly how aesthetics worked in relation to the natural world rather than in relation to art. Unsurprisingly, but also frustratingly, most of the work I was able to find focused on modern aesthetics. In terms of formal theories of aesthetics in nature, the eighteenth century was pivotal: it saw philosophers including Edmund Burke (1729–1797) and Immanuel Kant (1724–1804) articulating the idea of the 'sublime', the response of awe-mixed-with-fear when the human mind is met with the sight of something almost too large to comprehend – such as a mountain. But I found it very hard to find anything which helped to explain why the participants in the Burnet debate so often seemed to mention the beauty of mountains in the same breath as they described their *usefulness*.

Back then I spent a lot of my time working in the Cambridge University Library, itself a testament to changing aesthetic tastes, at least in architecture: it was designed by the same man who designed Bankside Power Station, now the Tate Modern, and you can tell. The main Reading Room is big and airy and both simultaneously silent and noisy in the way that only a library filled with readers shuffling their feet, turning their pages, and coughing discreetly can be. The pile of books by my elbow was becoming increasingly obscure when I opened a book by a Polish philosopher Władysław Tatarkiewicz (1886–1980), author of a three-volume *History of Aesthetics*.

Tatarkiewicz had worked his way through reams of early modern sources in order to detail the aesthetic principles underpinning people's appreciation of beauty during the period. He commented that one, which he called the 'Great Theory' of beauty, was dominant, and prized artworks and objects which were in proportion, regular and symmetrical. This is exactly the category of beauty against which Burnet found mountains so very much wanting: to him they were disordered, asymmetrical, not in proportion with the world around them and thus inherently ugly and unappealing. However, Tatarkiewicz also identified something which he termed the 'aptness theory' of beauty as being at play during the sixteenth and seventeenth centuries. According to this theory, the beauty or appeal of an object lay in the 'correspondence between the object and its purpose and nature'. In other words, how good was it at doing what it was made to do? Under this principle, a gold shield, however dazzling, could not be beautiful, since it would be too soft to protect the person using it and too heavy to be easily carried anyway. By the same token, a rubbish bin, however plain its design might be, could be beautiful if it was excellently designed for the purpose of containing rubbish.[44]

I sat back in my chair in that imposing reading room, surrounded by bespectacled lecturers, and resisted the urge to pump my fist in the air in a display of scholarly triumph which would surely have earned me a few reproving glances. Reading Tatarkiewicz's idea of the early modern 'aptness theory' of beauty was like finding the key to a hitherto-locked door and hearing it slot neatly into the barrel and releasing the bolt. The aptness theory of beauty was what underpinned so many of the arguments made against Burnet and in favour of mountains. Repeated statements that mountains were *useful*, that they fulfilled a *purpose*, were not weak protestations. They were statements that mountains were beautiful because they were useful. They were not merely utilitarian and ugly.

Nowhere is this more clearly articulated than in the published lectures of the classicist and theologian Richard Bentley (1662–1742). In 1692, he was appointed the first 'Boyle lecturer'. Robert Boyle (1627–1691), today considered a pioneer of modern chemistry and experimental science, had left money in his will to endow a series of lectures. These lectures were specifically intended to consider the relationship between natural philosophy (science) and the chief tenets of Christianity – the very same overwhelmingly important relationship which I highlighted at the beginning of this chapter. Richard Bentley started this series (which ran regularly until the start of the twentieth century and were reinaugurated as an annual event in 2004) with a bang: he delivered, across the course of eight lectures (or 'sermons'), a *Confutation of Atheism from the Origin and Frame of the World*. His eighth and final lecture opened with the assertion that 'the frame of the present world could neither be made nor preserved without the power of God' and that 'the order and beauty' of the world was the result not of chance but the intention of 'an intelligent and benign agent'.[45]

Thomas Burnet received special attention in the closing passages of this lecture. Bentley gave no names, but the educated listener would be under no illusions that he was referring to the *Theory of the Earth* when he commented scornfully:

> But some men are out of love with the features and mien of Earth; they do not like this rugged and irregular surface, these precipices and valleys and the gaping channel of the ocean. This with them is deformity, and rather carries the face of a ruin ... than a work of divine artifice. They would [wish] the vast body of a planet to be as elegant and round as a factitious [false] globe represents it; to be every where smooth and equable.[46]

Bentley rapidly made it clear that as far as he was concerned this was a piece of complete and utter nonsense, and his reasoning represented Tatarkiewicz's aptness theory dressed in different words:

> All pulchritude [beauty] is relative; and all bodies are truly and physically beautiful under all possible shapes and proportions; that are good in their kind, that are fit for their proper uses and ends of their natures.[47]

The mountains, therefore, should not be deemed 'misshapen' merely because 'they are not exact pyramids or cones'. In contrast, they should be valued as extremely beautiful because the irregularity they introduce into the world is of enormous benefit. Vapours (according to Bentley) condensed along their rocky sides, and thus 'give the plains and valleys themselves the fertility they boast of'. Different plants grow at different altitudes, so if the world was flat as Burnet wished it to be then 'we should lose a considerable share of the vegetable kingdom'. Finally, mountains were the source of metals, without which mankind would not have developed tools to harvest crops, to build cities, or (and here Bentley perhaps got carried away) to construct any sort of intelligent or advanced civilisation at all.[48]

What this makes clear is that it was not the case that early modern natural philosophers merely thought that mountains were useful *and* beautiful; they thought they were beautiful *because* they were useful. Even more importantly, they believed that usefulness to be the result of the deliberate design and intention of God. Mountains – jagged and disordered though they might have appeared to Thomas Burnet – were thus objects that put his contemporaries in mind of the most awe-inspiring thing imaginable to the seventeenth-century believer: the goodness and power of the divine.

Gloomy mountains?

Back when I first met Burnet, he inhabited an important role in the narrative that 'they didn't like mountains back then'. His one brief passage in praise of mountains which I quoted above – 'There is something august and stately in the air of these things that inspires the mind with great thoughts and passions' – has struck generations of historians deeply. He has been seen, in fact, as one of the first European writers to express a sense of the 'sublime in nature'.

As a formal aesthetic concept, the sublime captures the idea that an object being viewed possesses the quality of *greatness*: it is something massive, even incalculable. This greatness in turn results in the viewer experiencing a sense of awe and inspiration, of delight mixed with fear or astonishment. The term is old – dating back to ancient Greece – but the traditional narrative is that it came to be associated with nature during the eighteenth century, in the writings of authors such as Joseph Addison and Edmund Burke. The general implication is that it was only in modern times that people started to look at nature in general and mountains in particular as sublime, and to experience that sensation of awe-mixed-with-fear in their presence.[49]

At the same time – and this is a theme I will return to in the next chapter – historians, and humans in general, love to identify 'who was first', and there is a thrill to being able to say 'ooh, maybe *this person* was a bit earlier than everyone else!' (An obvious, modern-day example of this is the fascination with the possibility that Mallory and Irvine might have summited Everest before the first recorded ascent of Hillary and Tenzing.) This is exactly what has happened to Burnet, and interpretations of his observations regarding the 'greatest objects of nature'. He has been typified, in this *specific* extract, as one of the first people to express a 'modern' sense of the sublime in nature. His negative comments on mountains, in contrast, have been typified as being the norm for his time. In disliking mountains, he was traditional, conventional. In acknowledging that the sight of them could be inspiring, he was a pioneer.

I think this is the reverse of what really occurred. The responses to Burnet show that his momentary admiration of mountains was a perfectly normal sensation for a seventeenth-century Englishman; it was his denigration of them which set him apart from his peers. This isn't to say I don't think Burnet is important in the history of the development of the sublime, but it is not as straightforward as him being 'first'. I think it was exactly his *dis*like of mountains and his attempt to prove that God did not create them which enabled the development of the sublime.

When I first started my work on mountains, I took it for granted that the old narrative was correct – that 'mountain gloom' held sway up until the eighteenth century. What I was interested in was what that looked like: how did people express their dislike of mountains before the birth of the natural sublime and of modern mountaineering? So, I went looking for examples ... and I found

Thomas Coryate, and shielings, and Philippe de Champaigne's *Christ Healing the Blind*, and many more examples which showed no such thing. However, what I also found were a few very explicit examples of writers expressing their distaste or discomfort with mountains ... dating to the decades immediately following the publication of Burnet's *Theory of the Earth*.

One prominent example can be found in the writings of Daniel Defoe, whom we first met in Chapter 2 marvelling over the simple happiness of the cave-dwelling family on the slopes of Mam Tor. Over the course of his *Tour thro' the Whole Island of Great Britain* Defoe makes it pretty clear that he heartily detested mountains – even the relatively humble ones of northern England. The Lake District, now celebrated as one of the most beloved and bucolic of Britain's national parks, he termed 'the wildest, most barren and frightful' place that he had ever had the misfortune to see.[50] In the Peak District, he admired the grand edifice of Chatsworth House, but shook his head sadly at the mystery that 'any man who had a genius suitable to so magnificent a design ... would build it in such a place where the mountains insult the clouds'.[51] Defoe would certainly not have agreed with Richard Bentley as to the 'pulchritude' of mountains'.

Why, then, did he dislike mountains so much? One explanation could be that he was simply not a particularly hardy or adventurous traveller: crossing the moorland hills between Halifax and Leeds he complained that they were 'so steep, so rugged, and sometimes too so slippery' that they all but prevented the safe passage of carriages across them.[52] However, it is also evident that Defoe had read Burnet from a comment he made when peering into Poole's Hole, a cave in Derbyshire. This limestone cavern – a tourist attraction to this day – struck him as evidence of 'the great rupture of the Earth's crust or shell, according to Mr. Burnet's Theory'.[53]

Other travellers who had absorbed Burnet's ideas include John Dennis (1658–1734) and Joseph Addison (1672–1719), which is where things get interesting, for both men have been credited as early articulators of the 'sublime in nature'. John Dennis did not directly cite Burnet, but when travelling through the Alps in 1688 he observed, in language which echoed the *Sacred Theory*, that the mountains he viewed 'were not a creation, but formed by universal destruction ... they are not only vast, but horrid, hideous, ghastly ruins'.[54] Nevertheless, these 'ruins of the old world' also struck Dennis as the 'greatest wonders of the new'. Making his own crossing of 'Mount Aiguebelette', where Coryate had been forced to take to a chair, he experienced strong and contradictory emotions: 'a delightful horror, a terrible joy ... I was infinitely pleased, I trembled.'[55]

Meanwhile, Joseph Addison is (in the history of aesthetics, at least) famous for his 1712 essay on 'the pleasures of the imagination', in which he argued for the aesthetic category of 'the great'. The 'great' ranked above beauty in terms of the intensity of the response it prompted in the viewer, and even something 'terrible or offensive' could offer a 'mixture of delight in the very disgust it gives us' if of sufficient scale and impressiveness.[56] Seven years before, he had viewed the Alps from Lake Geneva and wrote that they 'fill[ed] the mind with an agreeable kind of horror, and form one of the most misshapen scenes in the world'.[57] With this kind of language, it should come as no surprise to learn that one of Addison's more obscure pieces of writing was a Latin ode composed in enthusiastic praise of 'that most famous man, Dr. Thomas Burnet'.[58]

What was going on here? My theory is this: before Thomas Burnet, most educated Europeans saw mountains as useful, beautiful and awe-inspiring. The most important aspect underpinning this response was the innate belief that God had made them. As Burnet himself said, their very size made one think 'of God and His greatness'. The unquestioned understanding that mountains were part of God's creation was the foundation of the positive seventeenth-century aesthetic response to them.

Burnet, however, set a charge which cracked that foundation. No, he said. Mountains were not the result of design, their irregularities were not planned. They were the result of the Flood and as such merely a memorial to mankind losing their way so far that God was forced to destroy all life on Earth excepting a single family and a breeding pair of every animal. Removing God from the equation, mountains were just ugly.

And so a writer like Daniel Defoe, who perhaps found mountainous landscapes uncomfortable and inconvenient anyway, came to them without any religious or intellectual framework upon which he could base any sort of positive response. But, at the very same time, men like John Dennis and Joseph Addison, wholeheartedly agreeing with Burnet's theories, went to the mountains and still found themselves impressed by them. Thus, they began to develop new aesthetic theories to explain the sense of awe they felt when looking at mountains. And so the sublime – which allowed for mountains to inspire wonder, but did not depend on the assumption that God had created them – slowly took root as the paradigm for appreciating wild nature which has survived to this day. My theory, therefore, is that the change marked by Burnet's theory was not from 'mountain gloom' to 'mountain glory', but from an appreciation of mountains that depended upon a belief in God, to one that did not.[59] Some who wrote of the sublime in mountains would of course think of the divine, but it was no longer essential: as religious belief waned, the atheist, too, could travel to the mountains and feel uplifted without any thought of a Creator.

There is a curious epilogue to Defoe's loudly expressed distaste for mountains. His *Tour* was a resounding success and was reissued multiple times in the decades following his death. Over the course of these new editions, and as the new aesthetic of the natural sublime spread its wings, Defoe's gloomy mountains were slowly rehabilitated. Where in the first edition the mountains of northern England were 'monstrous high' they had become 'of a stupendous Height' by the time the 1742 edition was issued.[60] By the time of the eighth edition in 1778 the newly written preface reflected on the apparent revolution in attitudes towards the mountainous regions of Britain as evidenced by the gulf between Defoe's opinions and those of the 'modern traveller'. According to the editor, Westmorland and Cumberland – today's beloved Lake District – were 'formerly considered as little better than barren and inhospitable deserts' but had since become 'objects of pleasure' to all who made the journey to them.[61]

This comment highlights how short memory can be when it comes to charting what our predecessors thought. Set against the sixteenth- and seventeenth-century writings and paintings I have already shared, Defoe's dislike of mountains is clearly not representative of the overall attitudes held in the preceding centuries. However, barely fifty years after the first edition of his *Tour* his attitudes came to represent the end point of 'mountain gloom' out of which modern appreciation grew. The mistake here is in thinking that historical change can only move in one direction: that before modern people liked mountains, premodern people did not. I say the answer is more complicated. Before Thomas Burnet, the general European view of mountains was a positive one, rooted in religious belief. After him, at least in English writings, there was a period of aesthetic ambivalence, as people tried to figure out how to enjoy the rugged, ruinous mountains that Burnet had brought to light. Their answer was the sublime.

Of course, the simpler narrative is more appealing for one important reason: it allows us to identify cherished modern literary figures and mountaineers as being the 'first' to love mountains. And, as the next and final chapter will show, an obsession with 'firsts' is precisely how a mistake – the idea that 'they didn't like mountains back then' – came to enter the history books. Even more than that, it has shaped our entire modern-day experience of mountain landscapes.

Vantage Point: But Who Was First?

Clinton Thomas Dent (1850–1912), surgeon and mountaineer, made no less than eighteen attempts to climb the Aiguille du Dru (3,754m) before he could finally claim that he had made the first ascent of it in September 1878. His efforts were later described as 'a long and protracted war', and the Dru as his 'greatest conquest'.[1] Dent climbed during what is now known as the 'silver age of Alpinism'. In the preceding 'golden age' members of the British Alpine club, along with their Swiss and French guides, had knocked the tops off of most of the largest and more well-known peaks of the Alps. This triumphant era came to a tragic end with the first high-profile mountaineering disaster: Edward Whymper's 1865 inaugural ascent of the Matterhorn, during which four climbers died. The silver age saw the first ascent of many lower – but often more challenging – peaks. By the early 1880s, the silver age was over and the high Alps, in terms of climbers being able to declare 'I was first', were more or less climbed out. Dent was among a group of British climbers who urged members of the Alpine Club to turn their attention to the Caucasus mountains. There, peaks still stood unconquered and the region had yet to be overtaken by the swarm of tourists who had, by the late nineteenth century, come to infest the Alps.

For all this, Dent was wryly aware of the self-deception involved in claiming 'firsts' in the mountains. In 1883, he read a paper before members of the Alpine Club, looking back on his experiences in the 1870s enjoying 'mountaineering in the old style', but also contrasting it with the state of the sport in the present day. He recalled a climb of the 'Portienhorn' (now known as the Portjengrat, 3,654m), noting with pleasure that 'No trace of previous travellers could be found in the mountain.' Of course, no trace did not mean they never existed, and I think Dent's tongue was firmly in his cheek when he continued:

> Doubtless the mythical and ubiquitous chamois hunter had been up before us, for at the time I write of the district was noted for chamois; but even if he had it makes no difference. We have found it, long since, necessary to look upon ascents made by chamois hunters as counting for nothing.[2]

The fiction of 'firsts', he went on to observe, had become even more stretched in recent years. He listed seven different ways in which 'a mountain in the present day can be the means of bringing glory and honour to many climbers'. Thus, 'A' climbs a mountain and earns the first ascent. 'B ascends it' in the first *recorded* ascent. 'C goes up it', in the first ascent 'from the other side'. 'D' goes up the way A did and down the way C did, thus claiming the first full traverse of the peak. 'E' gets lost, and 'scrambles up the wrong way', thus being able to claim the 'first ascent by the east-north-east arête'. Next, 'F' climbs it 'the ordinary way' but can claim the novelty of being the first Englishman, or of making the first ascent without guides. Finally, 'G' is 'dragged up by his guides', but actually makes the first *real* ascent, 'because all the others were ignorant of the topographical details, and G's peak is nearly one metre higher than any other point'.[3]

I have a great fondness for Dent. It takes a certain type of person to be able to recognise the ridiculous, as well as the sublime, in an activity in which one is deeply invested. The question of 'who was first?' is written into the DNA of mountaineering and, I believe, modernity itself. But as Dent recognised, and as the final chapter of this book will relate, the very idea is an invention that requires us to close our eyes and ears to a past that might well have been first *before* us.

Chapter Five

How a Myth Becomes History

In May 2015 I was nearing the end of my second year as a PhD student and I was in a town called Jasper, in the Canadian Rockies, for a mountain studies conference. The days before the conference I had gone hiking, singing nervously as I went to alert the resident brown bears of my presence. These mountains were strange and wonderful to me, different from any I had encountered in my European travels: monumental, their tops rimed with snow, their forests filled with larger and more dangerous wildlife than I had ever hiked amidst before. In the town itself, enormous moose lay on patches of grass beside the railway and casually stopped traffic as they made their slow way across the roads.

The conference itself was also a new experience. Back home at my own university in Scotland, I knew no one else who studied mountains. One evening my supervisor overheard me in the pub, talking about my research, and he laughed and said, in a tone of mock weariness, 'Oh Dawn, you're *always* talking about mountains.' In the Rockies, though, I had first chance to talk about mountains with other people who were not only also deeply interested in them, but in some cases extremely senior and distinguished experts in their corner of the mountain studies map, whether twentieth-century mountaineering, modern poetry or conservationism. There were no other historians working on anything earlier than the eighteenth century. Why would there be, when everyone there very well knew that no one liked mountains before that date?

After all, it was an idea that – at the time – was taken for granted in academic literature. The most famous book on the topic was Marjorie Hope Nicolson's

Mountain Gloom and Mountain Glory, which was first published in 1959. Its title – borrowed from the writings of the nineteenth-century art critic John Ruskin – gave the alleged shift in attitudes an unforgettable label: from premodern mountain gloom to modern mountain glory. Nicolson has been cited by historians throughout the decades in support of the idea that the modern enthusiasm for mountains is a unique phenomenon.[1] The idea can also be found, as that chat at my dad's retirement party showed, far outside the academic echo-chamber. Probably the most famous book about mountains and history to be published in recent decades is Robert MacFarlane's *Mountains of the Mind* which observes, early on, that before the modern era '[t]he notion barely existed … that wild landscape might hold any sort of appeal', and mountains were viewed as being 'aesthetically repellent'.[2] A 2018 *New York Times* article arguing against the idea of applying urban-style safety regulations to mountain landscapes observed, by way of historical background, that before the mid-eighteenth century, mountains 'were seen as landscapes of evil otherness'.[3]

The paper I gave in the Canadian Rockies set out my evidence and arguments for why I thought this idea, which so many people accepted as historical truth, was wrong. In some ways I was almost a stereotype: the upstart young scholar with a new idea, claiming I had seen something that all who had gone before me had missed. At the end, I lowered my notes and looked around the room and braced myself, ready to be taken down a peg or two. The response was mixed. Some were positive and enthusiastic: they were persuaded that the old narrative was flawed and were excited by the new vistas of research that could be opened up if we left it behind. Others were more critical. One grand old professor, an enthusiastic mountaineer himself, leaned back in his chair, looked at me narrowly through his glasses and asked whether I was absolutely sure that the examples I had shared were not simply 'the exceptions that prove the rule'.

Years have passed since then, and I like to think I have found and shared enough 'exceptions' for the rule to be well and truly disproven. I searched and searched and I could not find evidence to uphold the idea that the customary response to mountains in the early modern period was one of fear and distaste. Yes, people acknowledged that they were dangerous, sometimes found them tiring, or even found crossing a pass in deep snow to be cold and unpleasant. The same is true today. I would challenge you to find any text written by a modern mountaineer that does not also contain reference to danger or physical strain, pain, or discomfort. I have ultimately come to believe that the only way you could possibly go to the early modern sources I have read and conclude that they hated mountains was if you had already taken for granted that this was the case. It was not a conclusion that had been reached by the neutral study of premodern mountain responses.

Where then did the idea come from? I decided to go digging for its origin-point. Who were the first people to come out and say that 'they didn't like mountains back then', and why?

I have already shared the earliest expression of the idea that I have been able to find, when the late eighteenth-century editors of Daniel Defoe's *Tour* insisted that recent decades had seen the birth of a new appreciation for wild nature which had been absent in Defoe's original text. In the century and a half that followed, this fairly moderate statement would crystallise into an assertion of a stark break between premodern and modern attitudes to mountains. This occurred most influentially in the writings of two individuals: William Wordsworth, the famous Romantic poet, and Leslie Stephen, one of the first presidents of the Alpine Club.

First among poets

Imagine the scene. Your name is William Wordsworth. You are 74 years old. More than half a lifetime ago, you and your good friend Samuel Taylor Coleridge launched (some say) Romantic literature in Britain with your *Lyrical Ballads*, which you had identified at the time as nothing less than a 'spontaneous overflow of powerful feelings', a new kind of writing 'materially different' from the formulaic verses in favour at the time. Your poems will be known for centuries to come as some of the most famous expressions of humanity's appreciation for nature. You oversaw the 'discovery' of the Lake District as a region of outstanding natural beauty. You have also made it your beloved home for many years. And then you hear the news: plans are afoot to build a railway, cutting right into the heart of the mountains and lakes of your poetry and your home.

In the modern day readers of this book may have heard of a phenomenon called 'nimbyism'. 'NIMBY' is an acronym for 'not in my back yard', and 'nimbyism' is the practice of objecting to a development that will affect your immediate locality. When I was growing up, for a year or two the village in which my grandparents lived was taken over by hand-painted signs, stubbornly declaring 'no gravel pit here!' My dad relished pointing out the fact that many of the driveways bearing these signs were covered in gravel. The even greater irony was that one of the most beloved features of the local landscape is a nature reserve centred upon a lake which is in fact a flooded gravel pit.

William Wordsworth was a brilliant poet, but he also proved himself to be a classic nimby. He loved his Lakeland but would rather it did not become *too* accessible for other people to fall in love with it as well. The idea of a railway running from Kendal, near the edge of the Lake District, right up to Lake Windermere (which, incidentally, was only a few miles from his home at Rydal Mount) brought the elderly wordsmith out of retirement with much the same fire as Thomas Burnet's *Theory* had lit in the octogenarian Herbert Croft.

Over the course of 1844, then, he addressed several pieces of writing to the editor of the *Morning Post*. First, and aptly enough given the source of his own fame, he sent a sonnet, which opened with a protest against the 'rash assault' represented by the planned railway, and closed with a plea for leaving the 'beautiful romance / Of nature' in peace.[4] When he wrote a longer prose letter in early December 1844, he rejected the idea that this was any mere 'local' debate: rather, the question was 'one in which all persons of taste must be interested', hence addressing it for readers of a national paper.[5]

Wordsworth opened by acknowledging that the main argument made for the railway was that it would 'place the beauties of the Lake District within easier reach of those who cannot afford to pay for ordinary conveyances' such as coaches, and that the poor would be 'wronged' should it not be built.[6] Wordsworth's response to this claim was a subtle one: would 'the poor', he asked, actually benefit – in the sense of being personally edified by the natural beauties of the region – if they were offered cheaper and more convenient access to the Lakes? Essentially his question was whether poor people were even capable of appreciating the rugged landscape to which the railway would transport them.

In answering his own question Wordsworth invoked history. '[T]he relish for choice and picturesque natural *scenery*,' he claimed, 'is quite of recent origin.' He listed famous examples of seventeenth- and eighteenth-century travellers disdaining mountains. Examples included John Evelyn, who in his Alpine crossing 'dilates upon the terrible, the melancholy, and the uncomfortable'. Interestingly, he identified Thomas Burnet as an early outlier with his momentary praise of mountains. But with that one exception, Wordsworth insisted, 'there is not ... a single English traveller whose published writings would disprove the assertion, that, where precipitous rocks and mountains are mentioned at all, they are spoken of as objects of dislike and fear, not of admiration.' According to Wordsworth it would not be until the poet Thomas Gray wrote his ode at the Grand Chartreuse in 1741 that attitudes to mountains would begin to change.

This represents the earliest clear articulation of a supposed shift from 'mountain gloom' to 'mountain glory', from the pen of a geriatric poet determined to save his home from tourism. Wordsworth constructed his historical narrative in order to prove that the appreciation of 'romantic scenery' – mountains, rocks and lakes – was not innate, and instead 'must be gradually developed in both people and individuals'. To his credit, Wordsworth agreed that this taste, which had been growing among the 'educated' for some years, was worth cultivating more widely. However, he did not think that simply transporting hundreds of labourers and shopkeepers (his terms) to the Lakes was the way to go about it. Would it not be better, he proposed, to encourage them to regularly enjoy nature in the form of the fields just outside their hometowns? Weekly walks along hedgerows could be just the sort of training needed before this sort of person could enjoy 'a profitable intercourse with Nature where she is most distinguished by the majesty and sublimity of her forms'. But you could not just take someone straight from a gritty town centre to the banks of Windermere and expect them to really enjoy it. History proved that the taste for mountains had taken many years to develop at all; spreading it throughout society would be a similarly slow process.

Wordsworth closed his letter in a pessimistic spiral. Since the 'humbler classes' would not be ready to enjoy the quiet, sublime pleasures of the Lakes, the railway companies would be forced to 'devise or encourage entertainments' to tempt them across to Windermere. Thus, 'we should have wrestling matches, horse and boat races without number, and pot-houses and beer-shops would keep pace with these excitements and recreations.' Which, Wordsworth thought, was frankly pointless, since such entertainments could be had anywhere; they did not need to take place on the blessed banks of Lake Windermere. But if the railway did come, then the peace of the Sabbath day would be destroyed, and the beauty and retirement of the region shattered.

On the one hand this whole episode is almost comedic. I live in a pretty coastal village which is popular with holidaymakers and second-home owners alike, and the local newsletter periodically erupts into controversy over houses being bought up by people who only intend to use them at weekends or during the summers. Wordsworth's anxious, hyperbolic pleas against the railway are all too similar, and enormously human. He loved the Lake District, but one of the things he loved it for was that it was quiet and remote. His affection for it had to be selfish, because if too many other people came to love it as well then it would no longer be the same place. Like so many of us, the great poet did not really want to share his toys.

Reading between the lines, though, there is more to it than that. What was Wordsworth really saying, amidst his talk of railways, about the love of mountains and about his own relationship with them? He was saying that the love of mountains was a relatively new thing, and still a relatively rare thing, confined only to those with the refinement and education to have acquired it. He was saying that he, William Wordsworth, was one of those rare few. That he, a poet who inaugurated a new way of writing, was among the first to really love mountains.

The railway to Windermere opened, despite Wordsworth's impassioned objections, in 1847. I doubt whether he was very happy about this, but he may have been comforted by the assessments made by posterity regarding the origins of the 'modern' taste for nature. Over the late nineteenth and early twentieth centuries, numerous literary scholars set themselves the task of identifying which writer was the first to express an appreciation for mountains. J.C. Shairp, Professor of Poetry at the University of Oxford, traced the gradual ascent towards the modern appreciation of nature in his 1877 book *On the Poetic Interpretation of Nature*. Shakespeare, he observed, never mentioned mountains, and so 'the mountain rapture had to lie dumb for two more centuries before it found utterance in English song'.[7] Shairp gave credit to several poets, including Robert Burns, for opening up 'a new avenue of vision into the life of Nature', but the title of his final chapter made it clear that the pinnacle belonged to

'Wordsworth as an Interpreter of Nature'. It was Wordsworth who was the 'great leader' of the Romantic movement and who wrote of nature most 'truly, broadly, and penetratingly'.[8]

Half a decade later, the critic Edmund Gosse would agree instead with Wordsworth's own assessment that Thomas Gray was 'the first of the romantic lovers of nature', in contrast with 'the old-fashioned fear' of grand scenery.[9] In *The Development of Feeling for Nature* (1888) Alfred Biese, a German literary historian, wrote that the early modern period 'showed no trace of mountain admiration' with the solitary exception of Conrad Gessner. In terms of the modern era he barely mentioned Wordsworth, and neglected Gray entirely. Instead, he identified the French philosopher Jean-Jacques Rousseau as the person 'who first discovered that the Alps were beautiful', and 'the real exponent of rapture for the high Alps and romantic scenery in general'.[10]

These are just a few examples of the many attempts made by literary critics and scholars of the late nineteenth and early twentieth centuries to chart the origins of the modern 'feeling for nature' in general and for mountains in particular. What is interesting in all of this is that the question was never over whether the feeling for nature *was* modern. The question – taking that as a given – was always 'but who was really first?'

Mountaineers at the summit position

A quarter of a century after Wordsworth's attempt to block the construction of a railway line, the mountaineer and literary critic Leslie Stephen (1832–1904) was sitting on a train leaving London, in the company of a 'highly intelligent Swiss guide'. Looking out at the 'dreary expanse of chimney pots', Stephen observed to his travelling companion that the view was not as fine as the one the pair had shared from the top of Mont Blanc. 'Ah sir!' sighed the guide, 'it is far finer!'

Leslie Stephen is one of the few nineteenth-century historical figures who can be introduced by way of his relationship to a more famous woman: he was the father of Virginia Woolf, the novelist. Although he did not inaugurate a new form of literature, as his modernist daughter did, his own achievements were still worthy of record. He served as the founding editor of the *Dictionary of National Biography* which, until the advent of Wikipedia, was the go-to source for information about British individuals of note. He was also a highly successful mountaineer: he joined the Alpine Club in the year of its formation, served as its president from 1866 to 1868 and made the first recorded ascents of nine alpine peaks. In 1871 he published *The Playground of Europe*, which was mostly a collection of accounts of his various climbs. However, he opened his book with the scene on the train, a preamble to a lengthy essay on 'The Love of Mountain Scenery'.

In some ways it is a strange start to a book recounting high-altitude adventures. The essay fills almost seventy pages in the first edition and is loaded with literary quotations, references to Rousseau, and reflections on the relationship between science and poetry. Far from taking any direct route to a mountain summit it meanders between different examples. The essay is divided into two halves: 'The Old School' and 'The New School' which held sway before and after the birth of the modern love of mountain scenery.

His guide's preference for London chimney pots over the Alps was emblematic of 'the Old School'. Stephen pointed out that the shock he felt at his guide's words likewise 'await the student of early Alpine literature'. He gave a few examples, including Abraham Ruchat's *Délices de la Suisse* which we already encountered in Chapter 2 with its account of the adventures of the chamois hunter Stoëri. Stephen pointed out that Ruchat himself spoke disparagingly of the 'prodigious height and eternal snows' of the Alps. Instead, he and other 'early' authors on the Alps, according to Stephen, tended to comfort themselves by listing the useful resources which the mountains provided, such as crystals and 'fur-bearing animals' to be hunted. Stephen's commentary on this was that someone inclined to praise mountains on such grounds surely 'hated the mountains as a sea-sick traveller hates the ocean, though he may feebly remind himself that it is a good place to fish'.[11]

Among other famous literary figures, Stephen mused on the mountain experiences of Samuel Johnson, the stoutly built compiler of the famous *Dictionary of the English Language*. Johnson visited the Highlands in 1773, when – by Stephen's own account – the love of mountain scenery was already in the ascendant, but he had little positive to say other than that a good philosopher ought to be familiar with all types of landscape, including barren mountains. Stephen observed wryly that, 'it would be difficult to imagine a human being more thoroughly out of his element than Dr. Johnson on a mountain,' and excused him 'for expressing frankly sentiments which a considerable number of modern tourists might probably discover at the bottom of their hearts'.[12] It is here we get to the crux of it. Maybe the real question, Stephen wondered, shouldn't be why it was that people in the past didn't like mountains, but rather why did people today love them so?

Stephen's overarching perspective was, ultimately, very similar to Wordsworth's: that the love of mountains was not something innate but rather a taste which had, in relatively recent times, come to be developed. He rejected the idea that shifting attitudes to the Alps could be explained solely with reference to improvements in roads and ease of travel. Rather:

> [t]he mountains, like music, require not only the absence of disturbing causes, but the presence of a delicate and cultivated taste. Early travellers might perceive the same objects with their outward sense; but they were affected as a thoroughly unmusical person is affected by the notes of some complex harmony, as a chaos of unmeaning sounds.[13]

In other words, you had to have an 'ear' for mountains, and most premodern Europeans simply did not have it.

In the second half of his essay, 'The New School', Stephen went in search of the 'precise period' in which the love of mountains 'became a recognised and vigorous reality'.[14] He did not, alas for William, give the laurels to Wordsworth. Instead, he handed them to a Genevan philosopher, musing that, 'If [Jean-Jacques] Rousseau were tried for the crime of setting up mountains as objects of human worship, he would be tried by any impartial jury.'[15]

It was Rousseau's famous novel *Julie, ou la nouvelle Héloïse* that earned him this accolade. It was first published in 1761 and was originally titled 'Letters from two lovers, living in a small town at the foot of the Alps' (*Lettres de Deux Amans, Habitans d'une petite Ville au pied des Alpes*). The book was a runaway hit, the bestseller of the century. Early on in the epistolary exchange Julie's lover writes to her of his experience walking in the Alps, where he found that, 'the spectacle has something indescribably magical, supernatural about it that ravishes the spirit and the senses.'[16] So it was that a book about two star-crossed lovers took the love of mountain scenery and shared it with delighted readers across Europe.

It may seem surprising that Stephen, a mountaineer, should identify a novelist as the originator of the modern love of mountains. He was, however, quick to note that Rousseau had 'accomplices' in his crime. The most important of these was Horace Bénédict de Saussure (1740–1799). Another Genevan, it was Saussure who famously conceived of the idea of ascending Mont Blanc, the highest mountain in Europe. Saussure must have liked competition, for he offered a reward to the first person to reach the summit whilst continuing to make attempts himself. He was pipped to the post by Michel Paccard and Jacques Balmat in 1786, but a year later became the third recorded person on the top of Mont Blanc – a bronze medal, at least. Or, as Stephen put it, 'Saussure deserves the unfeigned reverence of every true mountaineer.'[17]

It was to Saussure that Stephen gave the honour of providing a clear date for the origin of a passion for mountains. 'The dividing line,' according to Stephen, 'may be drawn about 1760, and the Alps were fairly inaugurated … as a public playground.' In 1760, Saussure paid his first visit to Chamonix, and 1761 was the year he offered his reward. It would be claimed twenty-five years later, by which point the Alps had become a honeypot drawing travellers from far and wide. It was also in 1761 that Rousseau published his *Julie*. The key distinction between Rousseau and Saussure was that, whilst Rousseau 'showed the promised land distinctly', he 'did not himself enter into and possess it'. It was Saussure, the mountaineer, who took the steps into the physical mountain fastnesses to which Rousseau's fiction had lent such literary fame.[18]

Together, the two Genevans inaugurated a new way of seeing mountains, at least in Stephen's account. And what came before? 'We may say, then,' Stephen opined, 'that before the turning-point of the eighteenth century a civilised being might, if he pleased, regard the Alps with unmitigated horror.'[19]

Despite this strong statement, Stephen does admit to a few exceptions to his rule – all of which will, by this point, be familiar to you. Indeed, I think I could have a very interesting conversation with Stephen – perhaps climbing a hillside, rather than at the party I imagined for our seventeenth-century natural philosopher in Chapter 4 – about our respective interpretations of some of my historical 'friends'. He wrote of Scheuchzer and his dragons, and also of 'the learned Jesuit Kircher'. As well as visiting Etna and Vesuvius, Kircher also visited the Alps, where he was fascinated by tales of dragons and, I imagine, looked hopefully to the skies for the chance of seeing one himself.

'Old travellers saw a mountain and called it simply a hideous excrescence,' wrote Stephen, but then they filled them with fairy tales. The folk myths – monsters, demons, gnomes and dragons – which had apparently populated mountains in early modern Europe represented something which intrigued Stephen. He wondered whether they were 'merely expressing in another way the same sense of awe which we describe by calling the mountain itself sublime and beautiful?'[20] This is not a million miles away from my own interpretation of Scheuchzer's dragons, which is that stories of frightening beasts in the mountains partly reflected the real mountaineers' lived experience of the dangers of the high landscapes.

By the late seventeenth and early eighteenth centuries, however, 'dragons and goblins were ... at the fag end of their existence'. The death knell for the belief in mountain dragons, according to Stephen, coincided with the very same 'struggle between scepticism and faith' which I discussed at the beginning of Chapter 4. And, just like me, Stephen saw Thomas Burnet as playing an important role. He quoted Burnet's 'poetical passage' upon the august and stately air of mountains, and also his horrified imaginings of the mountains of the Earth when 'newborn and raw' as the floods subsided. By the end of the seventeenth century, the mountains were 'bare of imaginary beings', and reduced to 'big chaotic lumps' of stony matter.[21] Even the unsettling superstitions of fearful beings which had made them interesting – if not necessarily beautiful – were gone.

The Alpine Club was founded in 1857, and Leslie Stephen joined it a year later. The purpose of the club was, as you would think, to support members in climbing mountains, but it also did something else. It created a forum within which its members could tell stories. Often, these stories – shared in papers read to gatherings of club members or published in the pages of the *Alpine Journal* – relayed details of recent expeditions, of new peaks climbed. Sometimes, though, these were stories about the past. Members of the Alpine Club had a vested interest in exploring and retelling the history of how people had engaged with mountains because it helped them to identify and declare the place that their club, their sport, had within that history.

Understanding how the members of the Alpine Club viewed the history of mountains is not a straightforward task; the puzzle pieces are scattered throughout the early volumes of the *Alpine Journal*, and sometimes appear in unlikely locations. For example, it was in a review of an 1875 book about a modern journey across the Himalayas that I found the following:

> According to '*The Times*', this volume [Andrew Wilson's *The Abode of Snow*] is a record of 'systematic mountaineering', such as is seldom undertaken or described by members of the Alpine Club. We are sorry to see the leading journal expose both its complete ignorance of the subject it is talking about, and of the meaning of the words it uses. As we understand the word – and its introducers have perhaps the best right to define its meaning – Mr. Wilson's book is not a record of 'mountaineering' at all. It is the story of a journey made, with but one or two exceptions of a few miles, on horseback or on a litter. This kind of mountain travel was the only sort known to our ancestors. The modern passion for foot-climbing as an athletic sport was felt to be so distinct that a new word 'mountaineering,' had to be invented for it.[22]

This is a fascinating extract, putting aside the delightfully snooty put-down of *The Times* and its ignorance. The anonymous reviewer is simultaneously asserting the novelty of mountaineering as an entirely modern phenomena, dismissing all prior mountain travel as brief and rarely on foot, and most importantly insisting upon the right of mountaineers (such as himself) to define what mountaineering means. I wonder what kind of review this individual would make of this book, or of my suggestion that 'the real mountaineers' far predated the Alpine Club?

Another review gave even more insights into how members of the Alpine Club viewed their historic predecessors in the mountains. Rather surprisingly, this 1867 'review' was of a 1723 edition of Johann Jacob Scheuchzer's *Itinera Alpina*, first published in 1708. The reviewer is clear: 'Alpine travelling, in its modern acceptation, dates from but a very few years back.' The few, such as Scheuchzer, who had ventured into the Alps in earlier centuries 'had not yet learnt to appreciate the sublimities of the mountains'. They took an interest in nature, but in apparently small, even silly, things, such as 'a rock with a hole in it', reminding the reviewer of nothing so much as, 'the British cockney ... to whom Snowdon is nothing but a dirty and inconvenient mound [but who] will fall into ecstasies at a rock shaped like the late Duke of Wellington's nose.' From the perspective of the modern, scientifically minded mountaineer, writers like Scheuchzer 'remind us of children who still look upon nature as a great collection of quaint toys and ingenious puzzles'. All in all, the reviewer concludes that 'we have abundant proof that the beauty of the snowy Alps ... had not yet dawned upon travellers'.[23] The early moderns were children intrigued by trivialities but blind to the appeal of mountains.

The pages of the *Alpine Journal* did acknowledge some rare exceptions to this rule. Conrad Gessner was credited by one author in the journal as 'the morning star of mountaineering', and by another as 'about the earliest Alpine traveller who takes a real pleasure in the mountains for their own sake'.[24] An essay by one-time club president William Longman, published posthumously, observed of Gessner and his fellow Swiss authors that:

> The mountain passion was even then alive, smouldering on in the hearts of a few Swiss students for two centuries before it burst out in Savoy into the blaze which has now spread across Europe.[25]

Taken together, these observations make it very clear that Gessner et al. were seen as true exceptions – 'a few Swiss students' – rather than as evidence of any more widespread appreciation of mountains. If anything Gessner, 'the morning star of mountaineering', was almost inducted into the 'club' – a modern Alpinist displaced in time due to his apparently unusual love of mountains.

Later discussions of the history of mountaineering were rather more exclusive. Charles Edward Mathews read a paper on 'The Growth of Mountaineering' to mark the end of his time as club president, which was reprinted in that year's *Alpine Journal*. As an outgoing presidential address, it was laudatory; you can almost hear the back-patting. 'We are making history,' Mathews happily observed, 'with extraordinary rapidity.' However, 'the prehistoric epoch was not so long ago.' Here, Mathews didn't even mean the eighteenth century; he claimed that even as recently as 1855 mountaineering 'was in its earliest infancy'. What about Saussure, and Mont Blanc? It hardly counted to Mathews:

> Mont Blanc, of course, had been ascended many times, but that expedition is not necessarily mountaineering. The men who practised climbing for its own

sake were so few that they might be counted on the fingers of one hand, and the mountains that had then been ascended were not much more numerous than the climbers.

It was not, according to this narrative, until the 'original founders' of the Alpine Club began to visit the Alps, from 1854 onwards, that mountaineering truly began.

William Wordsworth undoubtedly had an agenda behind his gloomy depiction of premodern mountain attitudes: he wanted to prove that the taste for mountains was not innate and that as such it was pointless to build a railway to allow people who did not have that taste to access the mountains of the Lake District. Did Leslie Stephen have an agenda? I think he did, but it was more subtle, and maybe not one he was entirely conscious of himself. It is worth reminding ourselves where his essay on 'The Love of Mountain Scenery' appeared: as the opening chapter of a book dedicated to his own mountaineering adventures among the peaks and passes of the Alps.

The closing passages of the essay are particularly telling. Stephen is assured that his readers 'will agree that the love of mountains' – so recently attained – 'is intimately connected with all that is noblest in human nature'. More than this, he ended the essay and prefaced the rest of *The Playground of Europe* with the statement that:

> it should be ... the purpose of the following pages to prove that whilst all good and wise men necessarily love the mountains, those love them best who have wandered longest in their recesses, and have most endangered their own lives and those of their guides in the attempt to open out routes amongst them.[26]

What he was saying here was that it was mountaineers like himself who knew the mountains best and who loved them best. They were also – as the preceding essay had intended to prove – among the first to know them and love them. Stephen was not trying to make a polemical point like Wordsworth had been, but he was making a claim as real as any physical summit bid. He was claiming the summit of loving mountains first and best for the modern mountaineer.

A FEMALE PIONEER

William Wordsworth and Leslie Stephen represent two of the earliest articulations I have been able to find of the idea that 'they didn't like mountains back then'. However, by far the most famous version of the story can be found in Marjorie Hope Nicolson's 1959 *Mountain Gloom and Mountain Glory*. It was the very book which started me up the winding mountain paths of early modernity. The first time I read it I was intrigued. She laid out – in eloquent, readable detail – the context of premodern 'mountain gloom'. She made friends with Thomas Burnet decades before I did and unfolded how his *Theory* inaugurated the 'aesthetics of the infinite'. As she put it, it was only in the aftermath of his work that 'the "Mountain Glory" dawned, then shone full splendor'.[27]

In the first instance, I was entirely persuaded by her narrative. Indeed, I was fascinated by the idea of an era in which mountains inspired such different emotions from today, and was eager to learn more. Nicolson's key sources were texts from the early earth sciences – such as Burnet's *Theory* – and the writings of canonical figures from English literature. What, I wondered, did mountain gloom look like in other contexts, such as travel writing and art?

The research behind this book was my attempt to answer that question for myself. As I have already shown, my attempt to trace the wider contours of 'mountain gloom' found more exceptions to Nicolson's rule than I could count, and my admiration rapidly turned – with all the arrogance of youth – to irritation. How could this woman, clearly so brilliant, have missed so much? How could she have got it so wrong?

Older and a bit wiser as I came to write this book, I found myself wondering what the author of *Mountain Gloom and Mountain Glory* was really like. Beyond her published bibliography – which is extensive and impressive – I could find relatively little about Nicolson as a person beyond the fact that she was the first woman to hold a full professorship at Columbia University and that she was beloved by her students there, who called her 'Miss Nicky'. I wanted to go digging in archives to get to know her better, but at the time pandemic restrictions meant that an in-person trip was not possible. Fortunately, the generous scanning efforts of Smith College, Massachusetts, where Nicolson served as dean from 1929 to 1941, gave me some long-distance inkling of the excitement I had felt, aged 17, when meeting Sandy Irvine through his letters in Merton College, Oxford.[28]

I looked at photographs – a bonnie baby giving way to a toddler with a thoughtful and stubborn gleam in her eyes. Then a young woman, still looking introspective and determined. An outdoor shot, with her – wearing a white, drop-waisted, 1920s dress – encircled by craggy rocks that Thomas Burnet would have deemed unappealingly disordered, and looking at the person holding the camera as if to say, 'See? I told you I could get up here.' The image in which she looks happiest is one of her in profile and in full academic regalia: hood and gown. There are photographs too of her as an older woman at a function, possibly marking her retirement. In them she looks like the kind of partygoer who would be rewarding but a little intimidating to encounter, listening to the conversation around her with an intent and serious expression.

The documents I received – mostly tributes from colleagues and students, written either at the time of her retirement from Columbia, where she served as the first female professor of an Ivy League university, or at her death – conveyed a similar impression. 'Miss Nicky' was a brilliant teacher, unfailingly generous with her time and ideas, but also one who took no prisoners. As one tribute observed, the faith she showed in her students was one that was difficult to betray: if she saw potential in you, then you felt a duty to fulfil it. Having benefited from the tutelage of just such teachers in my own time, I felt that Nicolson would probably have filled me with a similar sense of affection mixed with slightly terrified awe.

In the opening of *Mountain Gloom and Mountain Glory*, she credited an undergraduate student, Aileen Ward, with first tending the roots of the tree that would become her monograph. I simply cannot imagine any modern-day academic opening a book with the acknowledgement that an undergraduate thesis contained 'the source of many of my ideas'. The respect and affection clearly went both ways: the dedication page of Ward's thesis bears the simple words 'For Marjorie Hope Nicolson'.

Such platitudes – that Nicolson was an inspiring teacher and a generous scholar – should not obscure the fact that she was also downright brilliant, something which her prose in *Mountain Gloom and Mountain Glory* had always led me to suspect. Early in her career, she was advised to avoid her topic of choice, metaphysical philosophy, since she neither drank whisky nor smoked a pipe: in other words, because she was a woman.²⁹ So she took English Literature, a far more acceptable area of study for a female, as her specialism … and used it to study metaphysical philosophy. She worked extensively on early modern ideas about the moon, writing with wry pleasure of seventeenth-century dreams of flying up to it. In *Mountain Gloom and Mountain Glory* she did something unprecedented. In Chapter 4, I wrote about how in the early modern period disciplinary boundaries as we now know them did not exist. In the mid-twentieth century, however, they most certainly did, but Nicolson confidently enfolded within the study of English literature the 'scientific' writings of the Burnet debate and more. Intellectually, just as with her appointment at Columbia, she crossed boundaries with aplomb.

Despite the photograph at the top of a crag, Nicolson has no name as a mountaineer, but she did reach several metaphorical summits first. I would not be here, writing a book about how people *did* like mountains back then, if she had not written her book stating so definitively that they *didn't*. Or, at least, I would be writing a very different type of book, because I have undoubtedly followed in her footsteps in making my history out of a synthesis of poetry and science, travel accounts and art. And, although she rejected the identity of being any sort of feminist pioneer – she far preferred to think of herself as a scholar, judged on the quality of her work rather than her sex – she also undoubtedly, with her lifted shoulders in her professorial robes, helped blaze the trail that saw me leaving my own doctoral graduation in a ridiculous, powder-blue gown surrounded by equal numbers of women and men dressed in the same attire.

This realisation was still a decade away when I wrote the earliest drafts of my PhD thesis, in which I made no bones of just how wrong I thought Nicolson had been. My supervisor was from Germany, where his role would be termed *Doktorvater*, or doctor-father. Occasionally he would give me the kind of advice that children find very hard to hear from parents. 'You keep saying Nicolson is *wrong*,' he said, shaking his head, 'how would you like it if someone said you were wrong?' The unspoken steer was that I could do with being less black-and-white.

Like most good parental advice, I must ultimately and grudgingly admit he was right – up to a point. I still think that the essential point which so many people have taken from *Mountain Gloom and Mountain Glory* – that people didn't like mountains before the eighteenth century – was wrong. But the interesting thing is that Nicolson made it clear from the outset that she did not invent the idea of premodern mountain gloom. On the very first page of the book she stated that her intention was to respond to 'a basic problem in the history of taste: Why did mountain attitudes change so spectacularly in England?' The idea that there *had* been a change was already thoroughly well known. *Mountain Gloom and Mountain Glory* just set out to explain it.

So, my belated response to my supervisor is to say that I now appreciate why Nicolson came to her conclusions: they were entirely reasonable ones for her to reach at the time. Decades of literary scholars – either consciously or unconsciously following in the footsteps of William Wordsworth – had repeated the old canard that the love of mountains was a relatively new feeling. In her introduction, Nicolson observed that, 'we see in Nature what we have been taught to see, we feel what we have been prepared to feel.'[30] The same, I would say, is true of history. She had been taught to see gloomy premodern attitudes to mountains, and so she found them.

Telling a different story: of mountains and modernity

It is important to remember that historians are themselves a part of history: we live in our own times, and the stories we end up telling about the past are inevitably influenced by the attitudes, politics and technologies of our own eras. I came to the topic of early modern mountains in an entirely different world from Marjorie Hope Nicolson. Nicolson set out to answer her 'basic problem in the history of taste' in an era in which the best available scholarly search tool was a card catalogue. If she wanted to read a rare early modern book she was reliant on finding a library which held a physical copy. I started my own investigation into early modern mountain attitudes in a digital age, with copies of every volume I could wish to read at my fingertips and the data-combing capacities of Google at my disposal. I am not saying that modern technology did all my work for me; just like Nicolson I still had to pore over the sources I had found and figure out how they all fitted together. But it did mean that the sources I started with inevitably encompassed a far greater number and range than Nicolson could have easily accessed. When I put them together, they formed a very different mosaic of mountain attitudes to the story she had told.

I was excited to share the new story I uncovered, not just with fellow researchers but also with mountaineers and the wider public. However, when I came to share it, I noticed something strange. What I had to say seemed to create a real frisson. When giving talks – such as the one in Canada – the body language of at least some people in the room would change. Arms would be folded, with shoulders tightly hunched. Questions were frequently uttered in a dubious tone.

For one memorable lecture, I had invited two friends along. One was training as a human-rights barrister, entirely accustomed to the chilly thrust-and-parry of the courtroom. Sitting in a nearby pub somewhat later in the evening, he raised his eyebrows over his pint and informed me that he had never heard anyone get such a grilling during questions.

The grilling continued online. In a particularly memorable response to one article I had written, a gentleman messaged me to say that though he welcomed my 'historical interest' into mountains, took issue with my profile photo:

> I cannot see what you have on your feet but the soles are white ... which above the tree level and on bare rock tells me heaps. You can wear bikini on the mountain top (as long as you have kit in your rucksack) but please wear a pair of boots, preferably with Vibram soles when anywhere but a river stroll.

For what it's worth, the photograph had been taken at a viewpoint five minutes' walk from a car park during a visit to the Appalachians. I was six months pregnant at the time, and tying laces had started to become troublesome, so I had chosen my footwear for the trip solely on the basis of them being easy to slip on and off. This aside, and once my initial indignation had settled down, I found it quite fascinating that my capacity to comment on mountains as a historian was being connected to my apparent lack of practical expertise as a mountaineer. I am not alone in this type of experience: one historian of modern mountaineering has shared an anecdote of visiting the Alpine Club to do research in their library and being asked how long he had been climbing. The idea that he could possibly have anything to say about the history of mountains without being an active climber was unthinkable.[31]

All in all, the response to my work has been a strange thing. Normally, historians don't upset people with their work. They might bore people, but they rarely make them feel threatened or angry. But it became rapidly clear to me that I was upsetting people. Was it just something about me? Was I, in truth, an insufficiently experienced mountaineer to have insights into the history of mountains? Or too busy having opinions whilst female, as that peculiar bikini comment seemed to imply?

In the end, I determined that it was more than that: the problem was not with who I was, but with what I was saying. And what, really, had I been saying? That the love of mountains is not unique to us in the modern era. That we did not invent it. But why was this such a sore spot?

At the Canadian mountains conference, I presented alongside a fellow historian named Peter Hansen, author of *The Summits of Modern Man: Mountaineering after the Enlightenment*. Hansen starts his book with the first ascent of Mont Blanc. As noted by Leslie Stephen, the (fairly slow) race to the summit was inaugurated by Horace Bénédict de Saussure (1740–1799), who was fascinated by geology and meteorology. His explorations of the Alps were, in some ways, intensely practical, motivated by the desire to take detailed and accurate sketches from high points, to carry out experiments on magnetism, and to take barometric readings. Yet he was also driven by the idea of reaching the summit of the highest mountain of Europe.[32]

In the end, de Saussure was not the first person to make that idea a reality. In 1786, Jacques Balmat (1762–1834) who earned his living hunting 'crystals' and rocks to sell to collectors and meteorologists, and Michel Gabriel Paccard (1757–1827), a doctor, reached the summit together. The reward offered by de Saussure in 1760 still stood and represented two weeks' earnings for Balmat. Both members of the pair had made previous attempts on the mountain, and both broke into a run when their goal was finally within reach.[33]

When they returned from their ascent, three things happened. Jacques Balmat learned that his newborn daughter had died only hours before he had reached the summit. Paccard took to his bed with snow blindness. And, almost immediately, people began to ask which of the two of them had really been the first to place his foot on the summit.

Peter Hansen calls this 'the summit position': the idea of the individual who was 'alone and first' on a mountain summit.[34] As Hansen points out, this is as much a myth as a reality, since mountain climbs are very rarely solitary. Even if one (or two) people stand on the summit, to get there they rely on the support of sponsors, porters, people who mapped part of the route before them, and so on. Hansen suggests that the idea of the summit position is written into the genetic code of both mountaineering and modernity itself.

⁂

'Modernity itself'. What does this mean? 'Modernity' is a term that I throw around with abandon, but it is also one whose meaning is not immediately obvious, or even fixed. It would be nice to be able to say, straightforwardly, that 'modernity' equates to a specific period of time: from, say, exactly 1750 to 1990. Unfortunately, whilst timelines might prefer to work that way, history does not always fit into such neat chronological boxes.

The problem is that in order to say that 'modernity started in year x', we need to be able to identify what 'makes' modernity. And there are a lot of different arguments for precisely what changes marked the shift to the modern era. Some link it to politics – the development of republicanism and ensuing disruptions of 'old world' orders such as the French Revolution. Others look at science – the shift from the multi-disciplinary, semi-theological pursuit described in Chapter 4 to the experimental method as we recognise it today. You can also look at technology, for example the growth of 'mass' communication, starting with Gutenberg's printing press in the fifteenth century all the way through the development of periodical newspapers in the eighteenth century and telegraphs and telephones in the nineteenth. Or you could think about social and family life, with 'modernity' marking a shift from multi-generational cohabitation to the nuclear family unit, and the gradual separation between home and work, with fewer and fewer 'cottage industries' and more and more workers making their daily way to factories or (depending upon their class) offices with the advent of the Industrial Revolution. The answer is really that 'modernity' is none of these things in isolation, but rather all of them taken together.[35]

This is all inevitably circular – 'modernity' refers to the changes which occurred during a particular timeframe, and the timeframe of modernity is defined by those changes, as identified by historians. Another big criticism of the idea of 'modernity' is that it is very much Western-centric. Some things identified as critical to modernity occurred long before in other societies – such as the development of movable type and the printing press, which took place around AD 1040 in China – but the date deemed critical for identifying the start of the modern age is still the publication of Gutenberg's Bible in Germany. Meanwhile, many other things that are deemed to mark the shift to the modern era are things which happened in the context of the European and American world. To take one example, it seems problematic to say that the nuclear family is 'modern' when many cultures around the world still see multiple generations living under the same roof.

At the same time, the past several centuries have seen the Western world doing its 'best' to impose its way of doing things around the world. Over the course of the nineteenth and twentieth centuries, the technology, science and social values of Western modernity were exported just as much as physical products. The world today is a far more homogenous and interconnected place than it was in the 1700s. It would probably be true to say that 'we are all modern now'. That said, some observers would argue that we are now living through a new, postmodern period of history. My response to that would be to say that I leave it to historians of 2200 to decide what to call our here and now. Even if we do end up falling into a different category for the purposes of studying and teaching history at universities, it is still undeniable that our experience of the world today is indelibly shaped by the melting pot of changes which made up modernity.

⁂

Modernity is also a way of thinking about ourselves within time. It is a way of thinking about history and where we belong within it. It is all too tempting to see the history of humanity as a history of progress, with 'us' – modern people – at the peak of it.

This way of thinking about time, history and progress has not always been the case. The Renaissance of the fifteenth and sixteenth centuries saw the rediscovery of classical texts which had been lost from view during the Middle Ages. Most educated Europeans during this time saw themselves as the fortunate inheritors of remnants of what had been the pinnacle of civilisation – ancient Greece – after long centuries of decline. For many scholars, it seemed that there was little they could do that would exceed or improve upon the thoughts and discoveries of 'the Ancients'.

However, the seventeenth century saw scholars questioning not just the literal truth of the Bible but also the pedestal upon which the Ancients had been placed. This formed the 'Quarrel of the Ancients and Moderns': the Ancients being the seventeenth-century scholars who saw themselves as 'dwarves standing on the shoulders of giants', whom they could never surpass, and the Moderns those who held that European thinking had moved and could continue to move far beyond the heights of classical antiquity. In the long run, the existence of 'the modern period' as a way of thinking about history would suggest that the Moderns won.

So, at the same time as William Wordsworth was helping to usher in Romanticism, and Leslie Stephen was enjoying the Golden Age of Alpinism, the whole idea of modernity was taking clearer and clearer form. This was the idea of modernity which insisted that we today, now, are doing things better than everyone did in the past. That we are the first to enjoy entirely new ways of thinking and doing things.

Peter Hansen has written that 'mountaineering and modernity mutually constituted one another'. By this he means that mountaineering, in the sense we now understand it, could not have developed before this idea of modernity. He is also saying that the very idea of modernity is epitomised and produced by the project of mountain climbing. Where mountaineering is focused on physically achieving 'the summit position', the metaphorical summit position is central to the idea of being modern. To be modern is to be at the top of the mountain of history: separate, different, better than our forefathers and foremothers.

Reading about the summit position and modernity, I finally understood why it was that my work evoked such strong reactions. At one level, it destabilises many of the stories that mountaineers have told themselves over the past two centuries about the origins of their beloved sport – that they were first not just to climb mountains but to love them. More generally, the new story I have uncovered threatens to undermine the summit position, a way of thinking which underpins the whole idea of being modern. Even if we are not consciously aware of it, 'being modern' is something we are all unquestionably invested in.

At several points over the past ten years, I have had to stop and ask myself why I feel the need to keep repeating my story when I have had ample evidence that it upsets and irritates many people who hear it. I also truly understand why it upsets them. Mountaineering is an enormous joy to the people who pursue it, and the history of the pursuit is an important part of it, the rocks at the base of the cairn of the modern experience of mountains. Who am I to remove one of those rocks, and threaten that joy?

And so, there's a part of me that thinks I had better keep 'wheesht', leaving the results of my research where they belong, in a dusty corner of my university's online repository for PhD theses, where very few people are likely to trip over it. Stop trying to pull down sacred cows. But I find myself returning to George Mallory's response when he was asked why he wanted to climb Mount Everest. 'Because it's there.' Why do I want to share all the stories of early modern Europeans who climbed mountains, admired them, painted them, wrote poems about them? Well, because they are there. Even more than that, the people to whom those stories belong have become my friends, and I am conscious of the obligation not to leave them dwelling in undeserved obscurity. I am not removing rocks from the cairn but revealing that the layers supporting it extend further down below the surface.

What really changed

The idea that 'they didn't like mountains back then' is a mistake that entered the history books in some truly fascinating ways. At each stage, the idea was a perfectly reasonable conclusion for each person who expressed it to come to. At the end of the day, that is because it is not a story about history or about how people in the past related to mountains. It is because it is a story about how we relate to mountains in modernity. It is a story which has stuck because it whispers something which is all too tempting to believe: that we were first to love mountains. We were first.

Except we were not the first. It was not the case that modernity saw the inauguration of an entirely new feeling for mountains. Overall distaste and fear did not transmute into love and awe. However, I do not intend anyone reading this book to put it down thinking that people in the early modern period related to mountains in *exactly the same ways* as we do today. Remember how I first came to decide to be an early modern historian: because of the strangeness of the era.

The indignant reviewer of Andrew Wilson's *Abode of Snow* was, to some extent, right. Modern mountaineering was something new – with a few qualifications. It did not invent the love of mountains. It did not even invent the basic techniques of crevasse-crossing or avalanche safety. What modern mountaineering did create was a sport and above all a sporting community. It has always been both a social and a physical pursuit. Gessner and Petrarch both spoke of the importance and pleasures of the company you keep on the hillside, and this formed part of the foundational make-up of modern mountaineering. The sport would not have developed into its modern form without the Alpine Club and all the many societies of climbers that followed rapidly on its heels. And these societies were not just for physical training; they were for climbers to meet one another, share stories, and to come together to plan their next great adventure. I cannot think of another modern physical pursuit quite like mountaineering. Has any other sport resulted in its enthusiasts spilling so much ink sharing their thoughts and experiences of it that they warrant whole book-length bibliographies, or countless runs of journals?[36]

During the early modern period, to climb a mountain did not in any way imply membership in a wider society of people engaged in the same pursuit. People also perceived and engaged with the whole space of the mountain differently, because the lure of the 'summit position' did not yet exist. An early modern traveller could afford to be vague, and say they crossed a particular mountain range rather than naming a specific mountain, because they gained no particular prestige by climbing one peak rather than another. They also did not necessarily climb all the way to the top; their high point might be dictated by where the best view could be gained, or where an object of some interest (such as a site of holy significance) might be found, or simply where the most convenient path was for herding cows along the edge of the hillside. Modern mountaineering brought a specific target and directionality to how we instinctively approach mountains today. You go all the way up them, and then back down. Variations may occur but only if, as Clinton Thomas Dent pointed out, they give you a new way of being 'first'.

Another genuine change is that modern mountaineering has come to dominate the mountains. What do I mean by this? During the early modern period, a whole range of different activities were associated with and left their marks upon the mountain landscape. Travellers, merchants and herders wore paths over the most convenient passes; shielings were built in the high pastures; botanists plucked samples from the hillsides. Today, most populated mountain regions are dominated by the activities of either mountaineering or skiing. They are also, I would wager, the activities most likely to come to mind were you to pluck a random person from the street and ask them, 'What sort of things do people do on mountains?'

Mountaineering has also changed which peaks we value most, or which ones come most easily to mind as significant or important mountains. In the early modern period, it was mostly mountains with classical associations which loomed largest: Parnassus, Mount Athos, Olympus. Today, these have been replaced in our cultural consciousness by Everest, K2, Kilimanjaro, and other very high or 'highest' mountains. It is mountaineering which has made the height of mountains matter above any other characteristic.

There have been other changes, too, which are not necessarily associated with mountaineering. Thomas Burnet's theories about mountains were revolutionary at the time but bear no resemblance to the geological sciences of today. When you step on a mountain today, you can think of it as the result of vast tectonic forces, which is an entirely different experience from thinking of it either as the creation of a benevolent God or as the end result of the Flood which had scoured most of humanity from the surface of the Earth.

The mountains themselves are different, too – shockingly so, given that only a few hundred years have passed. The results of climate change are such that it is entirely possible that the summer inhabitants of the alpage at Le Clou would struggle to recognise the view from their chapel if shown it today. Across mountain regions, glaciers have vanished, hiking traffic has worn ever wider and deeper scars across the hillsides, forests have been cut down or new plantations of fast-growing, non-native trees have been set to grow.

So there is much that separates the modern experience of mountains from that which came before. Yet still, for all that, some of the most deeply felt aspects are held in common. The novice climber today who realises that she is looking down on the clouds for the first time in her life is in some ways standing in the footprints of the Odcombian leg-stretcher, Thomas Coryate. The modern-day viewer of Philippe de Champaigne's *Christ Healing the Blind* will look at it and find their eyes drawn to the uplifting peaks just as the artist surely intended nearly 400 years ago. And regardless of religious beliefs, we might find ourselves agreeing wholeheartedly with the seventeenth-century natural philosophers who quoted the Bible in love and praise of the mountains. Whether we love to climb mountains or not, we too may be thankful for the chief things of the ancient mountains, and for the precious things of the lasting hills.[37]

Epilogue: Mountains After Mountaineering

In 1971, three Norwegians travelled to Rolwaling Valley in the Himalayas on what they termed an 'anti-expedition'. Arne Næss (1912–2009) was a philosopher who the year before had chained himself to rocks in front of a waterfall threatened by a dam-building project. Sigmund Kvaløy Setreng (1934–2014) was a one-time aircraft mechanic who studied philosophy under Næss and would later convert to Buddhism. Nils Farluund (born 1937) gave up a career in chemical engineering to become a professional mountaineer and guide and was an advocate for careful human immersion in 'free nature' – mountains, forests, the seas. The aim of the group was to *not* climb a mountain.

Instead, they spent several weeks living in the Sherpa community of Beding, at the foot of a mountain which they called Tseringma (7,034m, also known by its Nepali name, Gaurishankar). The trio set out to be as different as possible from the 'heavy, army-like expeditions' which had typified Himalayan mountaineering since the first Everest expeditions in the 1920s. These featured 'hundreds of ill-equipped porters' to carry the necessities and luxuries for privileged *sahibs*, and thoughtless pressure on the food supplies and workforces of small villages. They also made 'nationalistic, victory-driven attacks' on mountains that local inhabitants worshipped as sacred.[1]

The anti-expedition group, by contrast, travelled with only eight porters on their eight-day trek up the valley to Beding, largely to help carry food supplies which they had purchased in 'the abundance of the Kathmandu valley' rather than burden their hosts with further mouths to feed. They spent several weeks at Beding, both getting to know the culture of its inhabitants but also 'in physical and mental dialogue' with the mountain Tseringma. Sigmund spent most of his time in conversation with the Lama, learning more about Buddhism, and reaching the agreement that the climbers would not exceed an altitude of 6,000m upon the sacred peak.

Nils and Arne took two young men, Pasang and Lachpa, up into the mountain with them not as 'high-altitude porters', but as 'rope mates'. Nils 'ran a course in alpine climbing' for them on rocks near the village. At the time (and still, for the most part, today) Himalayan mountaineering had a heavy impact on the mountain environment, with ropes fixed to ice and rock using pitons, hundreds of porters, and bottles of pressurised oxygen which were generally discarded once empty. Nils's Alpine techniques, which avoided fixed ropes and favoured 'nuts' (metal wedges which could be threaded into a gap in a rock for temporary security and then removed), were far more 'nature friendly'.

Nils's lessons enabled the young men to be equals on the climbs – a far cry from George Mallory's vision of his porters as 'children' in their own mountains. For their part, Pasang and Lachpa shared their cooking of 'true, vegetarian Sherpa meals', though the group also feasted on a few Norwegian delicacies – fish and *geitost*, a sort of caramel-flavoured cheese. The group also took a six-day trek over a 5,755m pass to the neighbouring Khumbu valley to visit Pasang and Lachpa's home village. The goal was to get to know these 'rope mates', to teach them and learn from them, rather than use them as servants.

Shortly before this trek, Arne fell ill and decided to leave Beding. He took with him a petition to the King of Nepal, signed by all of the villagers, 'for a ban on permission for summit climbs on Tseringma and other holy Himalayan mountains according to Buddhist belief'. At the time, forty applications had been made from Western countries to the Nepali government to attempt a first ascent of Tseringma.[2] Permission was finally granted in 1979, and that same year an American–Nepalese expedition reached the highest summit. Tseringma is not a double-headed but a triple-headed mountain; its second top was summitted later the same year. The third and lowest top fell in 1981 to an Australian army expedition led by a major general and which travelled into base camp with 103 porters. The expedition report noted off-hand that 'our presence on the mountain was received with mixed feelings,' but the group of Western climbers 'returned to Australia satisfied'.[3] The holy mountain was conquered by the very opposite of the anti-expedition.

• • •

During the final stages of revising this book I went on my own anti-expedition, in a manner of speaking, whilst on holiday in the Lake District with my husband, children and parents. I have some personal sympathy for Wordsworth's irrational possessiveness for his Lakeland. It was a place we visited regularly during my childhood, and I cannot think of a landscape I love more in the world.

During our week there we were keen to climb 'a mountain', within the limitations set by the mileage of two small pairs of legs. We made for a hill called Latrigg, 368m high. A car park – from which we could also, with either fewer children or more foolhardiness, have climbed proud Skiddaw – offered us easy access to a trail less than a mile from the top.

My eldest, just turned 4, galloped ahead eagerly. We said she should stay on the path, rather than walking on the grass to either side, to avoid damaging the mountainside. She took this rule to heart and scolded us every time we stepped to the side to make way for fellow, faster walkers.

The youngest, not yet turned 2, stubbornly refused to go in a woven wrap on my back, instead demanding 'cuddles!' in my arms up the hill. 'Wrap or walk,' I insisted, and he stomped onwards, indignantly huffing, 'walk!' He would quite happily have stopped halfway up, for the path crossed a small stream which represented to his toddler self the greatest enjoyment he could possibly imagine.

There is a bench a little way below the top with views over Keswick which offered a similar distraction to his older sister. I reminded her that we would have a snack at the very top and off she ran again, her very first case of summit fever ignited by the promise of a chocolate bar. 'To the top, Mummy!' She enjoyed the view whilst we ate, optimistically misidentifying a large farmhouse as the holiday home we were staying in. The clouds had lifted from Skiddaw's double top, and I nodded a greeting to him.

The little one finally consented to being wrapped for the descent, though we then, contrary parents that we are, spent the descent trying to prevent him from taking an ill-timed nap on my back. Meanwhile the big girl noticed hanks of sheep's wool which had been shed by the herd which covered the hillside and which clung to plants along the side of the path. She filled her pockets with them, taking particular delight in an exceptional clump of black wool.

This was the only hilltop which we claimed during our week in Wordsworth's homeland, due to a combination of poor weather and the far greater attractions – as far as the small people were concerned – of playing in lakes and streams. It was my mother who missed the summit position most. One day, our feet in Derwentwater, she looked up at the mountains around us and sighed. 'I just want to be *up there*.' I looked around too, and had to admit that I didn't, really. I took joy in the sight of them, and I looked forward to the day when small legs had grown to the point that we could take longer hikes as a family. In that moment, though, I was as happy at the bottom of a mountain as I might have been at the top of one.

My early modern friends have certainly changed me. Fifteen years ago, I had my long-term route to the summit of Everest all mapped out. I had already gained some rope, snow and ice skills during a winter training week in Scotland. I knew I needed Alpine experience, too, and pored over adventure company brochures. I was particularly drawn to an expedition offered by one such company to the Tien Shan mountain range in Kyrgyzstan. The advertising material went to great lengths to emphasise that the region offered 'unclimbed' summits in Alp-sized packages: imagine being able to claim a first ascent in the same week as training in Alpine climbing methods! The next step would be seeing how my body responded at higher altitude, and I put together a shortlist of 7,000m peaks to potentially aim for.

I never even made it to Tien Shan. Life – finances, falling in love, getting married – intervened, and so did my early modern mountain research. By the time we took our honeymoon in the Alps, I was already looking at mountains differently. 'Because it's there' no longer seemed quite enough justification, on its own, for slogging to the top of a mountain. I paid more attention to the plants, animals and insects that I passed, wondering what Conrad Gessner would have said about them. I even found myself looking at the enormous rocks left behind by the receding glacier above Le Clou and thinking that maybe Thomas Burnet had a point.

Epilogue: Mountains After Mountaineering

❧ ☙

Being in the Lake District with my children drove this new perspective home for me. Instinctively, many children love to climb, to go *up* things. Even the toddler, once the option of being cuddled had been ruled out, needed no guidance to turn his rolling footsteps directly up the trail. Had we left them entirely to their own devices, though, I am not sure if they would have made it to the very top. They would have enjoyed the mountain in other ways, by throwing a hundred tiny stones into a stream, by collecting enough scraps of wool to make a jumper, or by hanging upside-down off the bench at the viewpoint. Like the anti-expedition group, children can show us what it means to slow down in a landscape.

Being with them on the hillside made me think about both the past and the future differently too. It brought home to me how much I have inevitably left out of these pages, how much I still don't know – is still unknown – about the experience of mountains in early modern times. Most of the research for this book was completed before my children arrived, and so I never even stopped to think that the 'real mountaineers' would have included the young as well as the fully grown. Did mothers of small babies at the shielings and hafods wear them whilst they worked, or lay them in baskets whilst they churned milk to cheese? Did they worry about nap timings? Did their 4-year-olds launch themselves up crags, nature's climbing frames? Did parents take advantage of the magpie-like nature of their youngsters and send them out to collect lichen for dyeing? At what age would a child first be shown how to cross a glacier safely? Like all historians, my research was shaped and limited by my own perspectives and experiences. There are so many more stories still to tell.

My daughter's rapid internalisation of our own concerns about erosion made me think of the future, too. One of the things which drove the members of the anti-expedition group was their anxiety regarding what they saw, in 1971, as a pending 'eco-catastrophe' promoted by a society focused on industrial-economic growth above all else.[4] Fifty years later, we are waist-deep in the ecological crisis which they foresaw, and the world our children will inherit is very different from the one in which we grew up.

I have to wonder what this means for the future of mountaineering and of travel to the mountains. It is a cliché for members of my generation to bemoan the ease with which our baby boomer parents were able to purchase their first homes. I wonder whether my children, in twenty years' time, will look with similar envy and astonishment at the financial and ethical ease with which millennials were able to explore the world. The route to Everest which I once saw for myself – pouring tonnes of CO2 into the atmosphere solely in the interest of my own individual sense of achievement – may well strike them as an indulgence they could never imagine for themselves.

There is an interesting phrase in modern mountaineering literature: climbing a mountain 'for its own sake'. I sometimes find myself wondering: what if the thing to do for the sake of a mountain is to not climb it?

This is a question which, I know, may well inspire a shudder of horror from some. How dare I suggest that it is wrong to climb mountains, just because I have lost the summit drive myself. On what grounds should those who love climbing mountains, who are good at it, give up that pleasure? I am conscious, too, of a sense of hypocrisy. Would I give up visiting the Lake District in order to preserve it? For I was guiltily aware during our stay of the immense pressure of the very mountain tourism in which I was participating. So many cars lined the side-streets of the town we were staying in that it was barely possible to drive through, let alone park. Perhaps Wordsworth was right to fear the crowds. The same is true of mountain regions across the world, and yet they have also grown to depend economically upon visitors to create jobs and provide local income.

The anti-expedition group 'were not opposed to climbing and summiting mountains in general'.[5] However, in avoiding any summit bid of Tseringma they modelled a different way of enjoying mountains, of connecting with a specific landscape and with the community inhabiting it. A Canadian writer and travel guide, Bob Henderson, is currently leading a legacy project to revisit Beding. 'We are not passing through quickly,' he says. 'We will get to know one valley well, not many valleys superficially.'[6] At the heart of the anti-expedition and its legacy project is the idea that there might be different ways to travel, different ways to appreciate mountains.

My book on early modern mountains does not have the answer to the environmental crisis, and the crisis goes far beyond mountains. There are innumerable considerations – reduction of industrial CO2 emissions, transitioning to electric-powered vehicles, cleaner energy – which are far beyond the remit of this historian to comment upon. But one small part of the puzzle may be figuring out new ways of exploring our world and enjoying nature.

In 2019, the Nepalese government, which charges over $10,000 apiece for climbing permits on Everest, sent an expedition to remove 11 tonnes of rubbish from the mountain. In May 2023, another Norwegian, Kristin Harila, broke records by climbing all fourteen of the world's highest mountains in less than four months. Her high-speed and high-profile ascents made use of helicopters to bring climbers to Base Camp (rather than having to trek in) and to deposit equipment at higher camps. On the one hand, this method, which damages both the environment and the employment prospects of Sherpas, has attracted criticism. On the other, it has been defended as being an 'increasingly common' approach to climbing in the Himalayas.[7] Further, more serious criticism followed when drone footage from K2, the final peak of Harila's record attempt, was released showing climbers walking past Mohammed Hassan, a dying porter. Harila insisted that her team tried to help him, but she – and many other climbers – nevertheless continued on to the summit.[8]

Edmund Whymper's disastrous ascent of the Matterhorn in 1865 marked the first ascent of that peak but also the first real critique of modern mountaineering, as onlookers realised that this athletic recreation carried mortal risks. Queen Victoria herself considered banning her subjects from climbing. It seems to me that the climate crisis in its turn brings into question the environmental sustainability of modern mountain climbing and tourism. Though it is a bitter pill to do so, it may be worth considering whether clearing rubbish from Everest and arguing about the use of helicopters goes even a fraction towards answering that question. A far greater shift in how we enjoy mountains in the decades and centuries to come may be needed.

At the elite level of mountaineering, the summit position has become increasingly difficult to obtain. In order for mountains to still be, as C.T. Dent put it over 100 years ago, 'the means of bringing glory and honour to many climbers', increasingly rarefied challenges have to be set. No longer is it possible to be first upon a high Alpine or Himalayan peak, or even to be the first to climb all of the mountains over 8,000m. Records are now set in terms of the youngest, the oldest, the fastest. The anti-expedition group would agree that the last thing needed today is a faster engagement with nature. The idea of 'slow tourism' already exists. What about 'slow mountaineering'?

And perhaps the early modern world, where no one ever flew by plane or thought of nature as a list of summits to be conquered, can help us imagine what that might look like. Those who travelled abroad were few and far between, and many people only ever experienced far and distant places through books illustrated with black-and-white engravings. Most people met the mountains and hills closer to home. Those who dwelt among the mountains lived off them, that is true, but they followed the rhythms of the seasons, taking more from the high pastures when they were rich with summer grass and leaving them to rest in winter. At times, mountains were rightly feared and treated with caution. It was entirely reasonable to step aside from the path to a summit if the way seemed too steep for you. You could, after all, enjoy the fine view to be found halfway up, or admire the way the mountain provided a habitat to a rich variety of plants and animals. Mountains were home, too, to old stories and mythical creatures. People looked at them and thought of the goodness of God. They evoked a sense of gratitude that they had been created.

Maybe to find a future for mountains we should look to the past.

Notes

Introduction

1 Daniel Lord Smail, *On Deep History and the Brain* (2008), p.3.
2 Kenneth Clark, 'The Worship of Nature', *Civilisation: A Personal View by Kenneth Clark*, episode 11, BBC (1969). Other examples of this narrative can be found in Claire Éliane Engel, *A History of Mountaineering in the Alps* (1950), pp.13–27; Francis Keenlyside, *Peaks and Pioneers: The Story of Mountaineering* (1975), pp.9–11; Fergus Fleming, *Killing Dragons: The Conquest of the Alps* (2000), pp.1–10.
3 Hermann Kirchner, 'An oration on travel', translated in Thomas Coryate, *Coryat's Crudities* (1611), fol. C6r.
4 These included Audrey Salkeld, *Last Climb* (1999) and Dave Hahn, Eric R. Simonson and Jochen Hemmleb, *Detectives on Everest* (2002).
5 The definitive and immensely readable biography of Irvine is Julie Summers, *Fearless on Everest: The Quest for Sandy Irvine* (2000).
6 Sandy Irvine, letter of 8 March 1924, The Sandy Irvine Archive, Merton College Oxford, Box 25/MEE/2.
7 Ibid., letter of 30 March 1924, Sandy Irvine Archive, Box 25/MEE/5.
8 Ibid., letter of 30 April 1924, Sandy Irvine Archive, Box 25/MEE/9.
9 United States Geographical Survey, 'What is the difference between "mountain", "hill", and "peak"; "lake" and "pond; or "river" and "creek?", www.usgs.gov/faqs/what-difference-between-mountain-hill-and-peak-lake-and-pond-or-river-and-creek
10 Select examples of published work on mountains in earlier periods include Gareth D. Williams, *Pietro Bembo on Etna: The Ascent of a*

Venetian Humanist (2017), Jason König, *The Folds of Olympus* (2022), and ed. Dawn Hollis and Jason König, *Mountain Dialogues from Antiquity to Modernity* (2021). The latter contains chapters by academics focusing on mountains in ancient, medieval, early modern, and modern contexts, and emphasising the continuity between them.

Vantage Point: Climbing Wine Barrels, Climbing Alps

1. Thomas Coryate, *Coryat's Crudities, hastily gobled up in five Moneths trauells* (1611), pp.486–9.
2. Ibid., p.69.
3. Ibid., p.70.
4. Ibid., p.70.
5. Ibid., pp.75–8.
6. Ibid., p.79.
7. Ibid., p.79.
8. Ibid., pp.79–80.
9. Ibid., p.80.
10. Ibid., p.476.

Chapter One: Mountain Ventures and Adventures

1. Jean de Thévenot, *The Travels of Monsieur de Thévenot into the Levant* (1687), p.168.
2. Ibid., p.168.
3. Ibid., p.167.
4. Ibid., p.169.
5. See Florian Cajori, 'History of Determinations of the Heights of Mountains', *Isis*, vol. 12 no. 3 (1929), pp.482–514.
6. Thévenot, *Travels*, p.171.
7. Ibid., p.171.
8. Ibid., p.194.
9. Ibid., p.193.
10. Ibid., p.169.
11. Ibid., p.171.
12. William Lithgow, *A most delectable, and true discourse, of an admired and painefull peregrination from Scotland, to the most famous kingdomes in Europe, Asia, and Affricke* (1614), sig. P3r-v. (Not all early modern books contain

page numbers, or do not feature them throughout. 'Sig.' is short for 'signature', which is a sequential code printed at the bottom of a page to help the book-binder fold and organise the pages correctly. 'r' and 'v' stand for 'recto' and 'verso', the front and back of a page.)

13 Clifford Edmund Bosworth, *An Intrepid Scot: William Lithgow of Lanark's Travels in the Ottoman Lands, North Africa and Central Europe,* 1609–1621 (2006), p.xvi.
14 Lithgow, *A most delectable and true discourse*, sig. P4v.
15 Ibid., G4r.
16 Lithgow, *The Total Discourse, of the rare aduentures, and painefull peregrinations of long nineteene years trauayles* (1632), p.390.
17 Lithgow, *The Pilgrimes Farewell, to his natiue countrey of Scotland* (1618), sig. Fr.
18 Adam Olearius, *The Voyages & Travels of the Ambassadors from the Duke of Holstein, to the Great Duke of Muscovy, and the King of Persia* (1662), p.159.
19 For Beş Barmaq and his Christmas Day ascent, see Olearius, *Voyages and Travels*, pp.205–6.
20 Olearius, *Voyages and Travels*, p.387.
21 Robert Davies (ed.), *The Life of Marmaduke Rawdon of York* (1863), pp.48–9. Though the biography of Rawdon was not printed until the nineteenth century, the original manuscript was compiled sometime in the seventeenth century.
22 Davies (ed.), *The Life of Marmaduke Rawdon*, p.50.
23 Ibid., p.51.
24 Ibid., pp.51–2.
25 Francesco Petrarch, letter to Father Dionighi da Borga San Sepolcro, 26 April 1336. I quote from the translation of the letter by Alan S. Weber in ed. and trans. Weber, *Because It's There: A Celebration of Mountaineering from 200 BC to Today* (Taylor Trade Publishing, 2003), pp.9–15.
26 Petrarch in *Because It's There*, p.9.
27 Ibid., p.10.
28 Ibid., p.11.
29 Ibid., p.13.
30 Peter H. Hansen, *The Summits of Modern Man: Mountaineering After the Enlightenment* (2013), pp.20–1.
31 Roger Frison-Roches and Sylvain Jouty, *A History of Mountain Climbing* (Flammarion, Paris & New York, 1996), p.20.

32 Douglas W. Freshfield, 'Placidus a Spesca and Early Mountaineering in the Bündner Oberland', *Alpine Journal*, vol. X (1880–1882), p.289.
33 These are both translated by Dan Hooley in Sean Ireton and Caroline Schaumann (editors), *Mountains and the German Mind: Translations from Gessner to Messner*, 1541–2009 (2020), pp.23–48.
34 Gessner, 'Letter to Jacob Vogel', in *Mountains and the German Mind*, p.30.
35 Ibid., 'A Description of Mt. Fractus (Pilatus)Ibid., pp.37–9.
36 Ibid., p.40.
37 C.T. Dent, *Mountaineering* (Badminton Library, London, 1892), pp.9–11.
38 Gessner, 'Letter to Jacob Vogel' in *Mountains and the German Mind*, pp.31–4.
39 Ibid., 'A Description of Mt. Fractus (Pilatus)' Ibid., p.37.
40 Ibid., p.40.
41 Ibid., p.41.

Chapter Two: The Real Mountaineers

1 Bernard Debarbieux and Gilles Rudaz, *The Mountain: A Political History from the Enlightenment to the Present*, trans. Jane Marie Todd (2015), pp.108–9 and pp.134–5.
2 An excellent insight into the British fascination with Tibet, before, during, and after the Everest attempts can be found in Peter Hopkirk, *Trespassers on the Roof of the World: The Race for Lhasa* (1982).
3 Letter from George Mallory to Ruth Mallory written from Base Camp, 9 June 1922.
4 Letter from George Mallory to Ruth Mallory from No. II Camp, 11 May 1924, relating to events of 4 and 5 May.
5 G.I. Finch, *Climbing Mount Everest* (1930), p.13.
6 Ibid., pp.27–8.
7 Letter from George Mallory to Ruth Mallory, 12 August 1921. This and previous transcribed with kind permission from the Master and Fellows of Magdalene College, Cambridge.
8 Dibyesh Anand, 'Western Colonial Representations of the Other: The Case of Exotica Tibet', in *New Political Science* vol. 29 (2007), pp.23–42.

9   Daniel Defoe, *Tour Thro' the Whole Island of Great Britain*, vol. 3 (1727), pp.46–9.
10 Ibid., pp.50–3.
11 Ibid., p.220 (note that vol. 3 has two 'paginations', or series of page numbers – the Scotland section starts anew at p.1).
12 Ibid., pp.225–6 (second pagination).
13 Thomas Kirke, *A Modern Account of Scotland* (1679).
14 Kirke, *A Modern Account of Scotland*, p.16. I explored the attitude of southern Englishmen towards northern England and Scotland in detail in 'The Contours of the North? British Mountains and Northern Peoples, 1600–1750' in eds. Dolly Jørgensen and Virginia Langum, *Visions of North in Premodern Europe* (2018), pp.223–42. The volume is available for free online.
15 William Cronon, 'The Trouble with Wilderness: or, Getting Back to the Wrong Nature', *Environmental History* vol. 1 no. 1 (Jan. 1996), pp.7–28.
16 Josias Simler, 'On the Difficulties and Dangers of Alpine Journeys', translated from the original Latin in Alan S. Weber, *Because It's There: A Celebration of Mountaineering, from 200 BC to Today* (2003), p.25.
17 On droving in Britain, see Richard J. Colyer, *The Welsh Cattle Drovers: Agriculture and the Welsh Cattle Trade before and during the Nineteenth Century* (1976), A.R.B. Haldane, *The Drove Roads of Scotland* (first published 1952, reissued 2011), and Ian Roberts, Richard Carlton and Alan Rushworth, *Drove Roads of Northumberland* (2010).
18 Simler, 'On the Difficulties and Dangers of Alpine Journeys', p.23.
19 Alan Cleaver and Lesley Park, *The Corpse Roads of Cumbria: Walks along the county's ancient paths* (2020) brings together fascinating historical details with guides to routes taking in traditional corpse roads.
20 Cleaver and Park, *Corpse Roads*, p.3.
21 Melissa Calaresu, 'Making and Eating Ice Cream in Naples: Rethinking Consumption in the Eighteenth Century', in *Past and Present* no. 220 (2013), pp.35–78.
22 Simler, 'On the Difficulties and Dangers of Alpine Journeys', pp.24–5.
23 John Ray, *Select Remains of the Learned John Ray* (London, 1760), pp.125–6. This volume saw the publication of his 'itineraries' as he recorded them in manuscript during his lifetime.
24 William Windham, *An Account of the Glacieres or Ice Alps in Savoy* (1744), pp.3–4.

25 Ruchat translated in Windham, *An Account of the Glacieres*, p.6.
26 Simler, 'On the Difficulties and Dangers of Alpine Journeys', p.25.
27 A study disproving the myth has been carried out: Benjamin Reuter and Jürg Schweizer, 'Avalanche triggering by sound: myth and truth', International Snow Science Workshop (Davos 2009), https://arc.lib.montana.edu/snow-science/item/252
28 Simler, 'On the Difficulties and Dangers of Alpine Journeys', pp.25–6.
29 Ibid., pp.26–7.
30 Ruchat translated in Windham, *An Account of the Glacieres*, p.6.
31 Simler, 'On the Difficulties and Dangers of Alpine Journeys', p.27.
32 Ibid., p.26.
33 Ruchat translated in Windham, *An Account of the Glacieres*, p.7.
34 Ibid., pp.9–10.
35 Simler, 'On the Difficulties and Dangers of Alpine Journeys', p.23; Ruchat translated in Windham, *An Account of the Glacieres*, p.9.
36 Simler, *loc. cit.*, Ruchat, *loc. cit.*
37 Simler, 'On the Difficulties and Dangers of Alpine Journeys', p.25.
38 Ibid., p.27.
39 Ibid., p.23.
40 Ibid., p.23.
41 Ibid., p.27.
42 Jesper Larsson, 'Labor division in an upland economy: workforce in a seventeenth-century transhumance system', *The History of the Family*, vol. 19 no. 3 (2014), pp.393–410.
43 Albert Bil, *The Shieling, 1600–1840: The Case of the Central Scottish Highlands* (John Donald Publishers Ltd, 1990), pp.200–201.
44 Ibid., pp.128–31.
45 John Mitchell, *The Shielings and Drove Ways of Loch Lomondside* (The Monument Press, 2000) and Donald MacDonald, 'Lewis Shielings', in *Review of Scottish Culture* no. 1 (1984), pp.29–33.
46 Scottish Archaeological Research Framework, 'Case Study: Transhumance and Shielings', ScARF National Framework (Modern Panel Report, September 2012). https://scarf.scot/national/scarf-modern-panel-report/modern-case-studies/case-study-transhumance-and-shielings/

Vantage Point: Healing the Blind

1 This information kindly provided by Glyn Hughes, Honorary Archivist at the Alpine Club.

Chapter Three: The Meanings of Mountains

1 Carl Sauer, 'The Morphology of Landscape', first published 1925, reprinted in *Land and Life: A Selection from the Writings of Carl Otwin Sauer* ed., J. Leighley (1963), pp.321–3.
2 W.G. Hoskins, *The Making of the English Landscape* (1955), throughout, p.233 for a particularly vivid example.
3 Denis Cosgrove, *Social Formation and Symbolic Landscape* (1984), p.13.
4 Thomas Churchyard, *The Worthines of Wales* (1587), sig. Mr-M3r.
5 Theuerdank was originally published in German in 1517. The summaries that follow are drawn by the modern edition by Stephan Füssel, *Theuerdank: The Epic of the Last Knight* (Taschen, 2018).
6 John Higgins, *The Mirror for Magistrates* (1587), fol. 55r.
7 Ibid., fol. 57r-58v. For further analysis of Brennus' mountain adventures, see Harriet Archer, 'Mountains, identity and the legend of King Brennus in the early modern imaginary', in eds. Dawn Hollis and Jason König, *Mountain Dialogues from Antiquity to Modernity* (2021), pp.197–214.
8 Leslie Stephen, *The Playground of Europe* (1871), p.63.
9 Marjorie Hope Nicolson, *Mountain Gloom and Mountain Glory* (1959), p.17.
10 The characterisation of the Aeneid as 'fanfiction' is not itself an original opinion; see for example Sadie MacDonald, 'In Defense of Fanfiction: Authors as Fanfic Writers', https://themindfulrambler.ca/2018/09/14/in-defense-of-fanfiction-authors-as-fanfic-writers/ accessed 10.05.2021.
11 Joshua Poole, *The English Parnassus* (London, 1657), sigs. A5r and A6v.
12 Ibid., p.137.
13 Ibid., p.233.
14 Ibid., p.275.
15 Ibid., p.352.
16 Ibid., p.440.
17 Ibid., p.473 and p.440.

18 Ibid., p.411.
19 Ibid., p.428.
20 Ibid., pp.283–4.
21 Ibid., p.540.
22 Ibid., p.313.
23 Ibid., p.416.
24 Ibid., p.297.
25 Ibid., pp.553–4.
26 Ibid., pp.343–4.
27 Ibid., pp.266–8.
28 Ibid., p.398.
29 Ibid., p.399 and p.403.
30 Ibid., p.407.
31 'Albion's Glorious Ile, the 400-year old colouring book – in pictures', *The Guardian*, 21 May 2016, www.theguardian.com/books/gallery/2016/may/21/albions-glorious-ile-the-400-year-old-colouring-book-in-pictures.
32 https://folklorethursday.com/legends/albions-glorious-ile-william-hole-and-the-strangest-maps-of-britain-ever-made/
33 Michael Drayton, *Poly-Olbion* vol. 2 (1622), pp.139–50.
34 Ibid., pp.131–2.
35 Skiddaw's ode to himself can be found in Drayton, *Poly-Olbion*, vol. 2 (1622), pp.165–6.

Vantage Point: Into the Volcano

1 Paula Findlen, *Athanasius Kircher: The Last Man Who Knew Everything* (2004).
2 Kircher translated in *The Vulcano's or, Burning and Fire-vomiting Mountains famous in the world* (London, 1669), p.49.
3 Ibid., pp.34–5.
4 I explore Kircher's response to Etna in more detail in 'Aesthetic Experience, Investigation and Classic Ground: Responses to Etna from the First Century CE to 1773', *The Journal of the Warburg and Courtauld Institutes* 83:1 (2020), pp.299–325.

Chapter Four: Mysteries of Science, Mysteries of Faith

1. Sir Thomas Pope Blount, *A Natural History: Containing Many not Common Observations: Extracted out of the Best Modern Writers* (London, 1693), Preface, sig. A4r.
2. Stephen Toulmin and June Goodfield, *The Discovery of Time* (1965), pp.87–8.
3. Paolo Rossi, *The Dark Abyss of Time: The History of the Earth and the History of Nations from Hooke to Vico*, translated by Lydia Cochrane (1984), p. ix.
4. Thomas Burnet, *The Theory of the Earth*, vol. 1 (1684), p.9. Throughout I quote from Burnet's English translation/version of the original *Telluris Theoria Sacra* (the Sacred Theory of the Earth), the first volume of which was published in 1681.
5. Burnet, *Theory*, vol. 1, pp.11–13 and pp.19–20.
6. Ibid., vol. 1, p.140.
7. Ibid., p.61.
8. Ibid., pp.67–8.
9. Ibid., p.68 and pp.71–6.
10. Ibid., p.77.
11. Ibid., p.110.
12. Burnet, *Theory*, vol. 2, pp.56–62.
13. Ibid., p.111.
14. Ibid., pp.135–8.
15. Burnet, *Theory*, vol. 1, pp.139–40.
16. Erasmus Warren, *Geologia* (1690), A2v.
17. Ibid., p.143.
18. Ibid., pp.145–6.
19. Ibid., pp.146–7.
20. Ibid., A2r.
21. Thomas Burnet, *An Answer to the Late Exceptions made by Mr Erasmus Warren against the Theory of the Earth* (1690), p.1.
22. Erasmus Warren, *A Defence of the Discourse Concerning the Earth Before the Flood* (1691), pp.31–2.
23. Thomas Burnet, *A Short Consideration of Mr Erasmus Warren's Defence of his Exceptions Against the Theory of the Earth* (1691), p.1.
24. Ibid., p.53.

25 Erasmus Warren, *Some Reflections upon the Short Consideration of the Defence of the Exceptions Against the Theory of the Earth* (1693), p.53.
26 Thomas Burnet, *An Answer to the Late Exceptions*, p.1.
27 Herbert Croft, *Some Animadversions Upon the Book Intituled the Theory of the Earth* (1685), sig. b1v.
28 Ibid., sig. b2r-v.
29 Ibid., sig. br.
30 Archibald Lovell, *A Summary of Material Heads Which May be Enlarged and Improved into a Compleat Answer to Dr. Burnet's Theory of the Earth* (London, 1696), p.14.
31 Croft, *Some Animadversions*, p.136.
32 Ibid., pp.136–7.
33 Ibid., pp.139–40.
34 Ibid., p.142.
35 Ibid., p.143.
36 Michel Foucault, *The Order of Things: An archaeology of the human sciences* (Routledge, 2002, first published in English 1970, first published 1966), pp.23–6.
37 For criticism of Foucault see for example Ian MacLean, 'Foucault's Renaissance Episteme Reassessed: An Aristotelian Counterblast', *Journal of the History of Ideas* (1998), vol. 59, no. 1, pp.149–66.
38 John Beaumont, *Considerations on a book, entituled the Theory of the Earth* (1693), pp.56–7.
39 Matthew Mackaile, *Terræ Prodromus Theoricus* (1691), sig. 2v. Mackaile's diction and spelling is so eccentric, even for the seventeenth century, that I have made him an exception in modernising my transcriptions.
40 Ibid., pp.3–4.
41 Ibid., p.22.
42 Ibid., p.26.
43 Beaumont, *Considerations*, pp.57–9.
44 Władysław Tatarkiewicz, *A History of Six Ideas: An Essay in Aesthetics* (1980), pp.125–33
45 Richard Bentley, *A Confutation of Atheism from the Origin and Frame of the World. The Third and Last Part ... Being the Eighth of the Lecture Founded by the Honourable Robert Boyle, Esquire* (London, 1693), p.32.
46 Ibid., p.32.

47 Ibid., p.37.
48 Ibid., pp.37–8.
49 For the ancient roots of the sublime, see James I. Porter, *The Sublime in Antiquity* (2016). For the more recent history of the sublime, see ed. Cian Duffy, *The Cambridge Companion to the Romantic Sublime* (2023).
50 Defoe, *Tour*, vol. 3., pp.223–4.
51 Ibid., pp.70–1.
52 Ibid., pp.113–4.
53 Ibid., p.59.
54 John Dennis, *Miscellanies in Verse and Prose* (London, 1683), p.139.
55 Ibid., pp.133–4.
56 Joseph Addison in *The Spectator* no. 412, Monday, 23 June 1712.
57 Joseph Addison, *Remarks on Several Parts of Italy, &c. in the Years 1701, 1702, 1703* (1705), p.455.
58 Joseph Addison, 'Ad Insignissimo Virum D. Tho. Burnettum, Sacræ Theoriæ Telluris Autorem', in *Musarum Anglicarum Analecta*, vol. 2 (1699), pp.284–6.
59 I make this argument in greater depth in Dawn Hollis, 'The natural sublime in the seventeenth century' in ed. Cian Duffy, *The Cambridge Companion to the Romantic Sublime* (2023), pp.29–40.
60 Quoted in Betty A. Schellenberg, 'Coterie Culture, the Print Trade, and the Emergence of the Lakes Tour, 1724-1787', *Eighteenth-Century Studies*, 44:2 (2011), p.207.
61 Ibid., p.216.

Vantage Point: But Who Was First?

1 Claire Éliane Engel, *A History of Mountaineering in the Alps* (1950), p.153.
2 C.T. Dent, 'Mountaineering in the Old Style', *Alpine Journal*, vol. II (1882–1884), p.393.
3 Ibid., pp.393–4.

Chapter Five: How a Myth Becomes History

1 Examples include Keith Thomas, *Man and the Natural World: Changing Attitudes in England, 1500–1800* (1983), pp.258–60 and p.290; Allen Carlson, *Aesthetics and the Environment: The Appreciation of Nature, Art,*

and Architecture (2000), p.72 and pp.83–5; Noah Heringham, *Romantic Rocks, Aesthetic Geology* (2004), pp.83–5; and Maurice Isserman and Stewart Weaver, *Fallen Giants: A History of Himalayan Mountaineering from the Age of Empire to the Age of Extremes* (2008), p.27 and p.457.
2 Robert MacFarlane, *Mountains of the Mind* (2003), p.14.
3 Francis Sanzaro, 'Keep Our Mountains Free. And Dangerous', *The New York Times*, 13 January 2018, www.nytimes.com/2018/01/13/opinion/sunday/keep-our-mountains-free-and-dangerous.html accessed 30 January 2023
4 William Wordsworth, 'Sonnet on the projected Kendal and Windermere railway', dated 12 October 1844, reproduced in *The Prose Works of William Wordsworth*, ed. Rev. Alexander B. Grosart, 3 vols., vol. 2 Aesthetical and Literary (1876), p.323.
5 Wordsworth, letter to the editor of the *Morning Post*, 9 December 1844, in *Prose Works of William Wordsworth*, vol. 2, p.325.
6 Grosart (ed.), *Prose Works of William Wordsworth*, vol. 2, p.326.
7 J.C. Shairp, *On Poetic Interpretation of Nature* (1877), p.170.
8 Ibid., p.234 and p.279.
9 Edmund Gosse, Gray (1882), p.32.
10 Alfred Biese, *The Development of the Feeling for Nature in the Middle Ages and Modern Times* (1905 English translation; original German edition 1888), p.261–4 and p.266.
11 Stephen, *The Playground of Europe*, pp.1–3.
12 Ibid., pp.9–10.
13 Ibid., p.14.
14 Ibid., p.35.
15 Ibid., p.38.
16 Jean-Jacques Rousseau, *Julie, or the New Heloise: Letters of Two Lovers who Live in a Small Town at the Foot of the Alps*, trans. Philip Stewart and Jean Vaché (The Collected Writings of Rousseau, vol. 6, 1997), p.65.
17 Stephen, *The Playground of Europe*, p.49.
18 Ibid., pp.30–1.
19 Ibid., pp.42–2.
20 Ibid., p.21.
21 Ibid., pp.2430.

22 Anonymous review of *The Abode of Snow* by Andrew Wilson (1875), *Alpine Journal* vol. VII (1874–1876), p.338.
23 'Scheuchzer's Itinera Alpina', *Alpine Journal* vol. III (1867), pp.200–5.
24 Douglas W. Freshfield, 'Placidus a Spesca and Early Mountaineering in the Bündner Oberland', *Alpine Journal*, vol. X (1880–1882), p.289, and Frederick Pollock, 'The Library of the Alpine Club', *Alpine Journal* vol. XII (1884–1886), p.428.
25 William Longman, 'Modern Mountaineering and the History of the Alpine Club', appended to *Alpine Journal* vol. VIII (1878), p.40.
26 Stephen, *The Playground of Europe*, p.68.
27 Marjorie Hope Nicolson, *Mountain Gloom and Mountain Glory: The Development of the Aesthetics of the Infinite* (1963; first published 1959), p.3.
28 Digitised materials relating to Nicolson can be found at https://findingaids.smith.edu/repositories/4/resources/349.
29 Andrea Walton, '"Scholar," "Lady," "Best Man in the English Department"? Recalling the Career of Marjorie Hope Nicolson', *History of Education Quarterly*, 40:2 (2000), p.176.
30 Ibid., p.1. I have traced the same development of the idea of mountain gloom and glory, from Wordsworth and Stephens to Nicolson, in 'Mountain gloom and mountain glory: the genealogy of an idea', *ISLE: Interdisciplinary Studies in Literature and Environment*, 26:4 (2019), pp.1038–61.
31 Paul Gilchrist, 'Gender and British Climbing Histories: Introduction', *Sport in History* 33:3 (2013), p.231.
32 Peter H. Hansen, *The Summits of Modern Man: Mountaineering After the Enlightenment* (Harvard University Press, 2013), p.63.
33 Ibid., p.89.
34 Ibid., pp.2–3 and p.11.
35 The classic article on modernity is S.N. Eisenstadt, 'Multiple Modernities', *Daedalus*, 129:1 (2000), pp.1–29.
36 See for example Jill Neate, *Mountaineering Literature: A Bibliography of Material Published in English* (1986).
37 Deuteronomy 33:15, quoted by Richard Bentley [etc].

Epilogue: Mountains After Mountaineering

1. Nils Farluund, 'The Tseringma Pilgrimage 1921 – an eco-philosophic "anti-expedition"', posted 14 August 2021, www.norgeshogfjellskole.no/the-tseringma-pilgrimage-1971-an-eco-philosophic-anti-expedition/
2. Bob Henderson, 'The story of the mountaineering anti-expedition of 1971', *Adventure Uncovered* published online 9 September 2021 https://adventureuncovered.com/stories/the-story-of-the-mountaineering-anti-expedition-of-1971/
3. P. A. Cullinan and G. Brammer, 'Australian Gauri Shankar (Tseringma) Expedition', *Himalayan Journal* 37 (1981), archived online at: www.himalayanclub.org/hj/37/6/australian-gauri-shankar-tseringma-expedition/
4. Bob Henderson, 'The story of the mountaineering anti-expedition of 1971', *Adventure Uncovered*.
5. Bob Henderson, 'Rolwaling Legacy project – 1971–2021 (2023). https:/bobhenderson.ca/1971-2021-anti-expedition/
6. Bob Henderson, 'The story of the mountaineering anti-expedition of 1971', *Adventure Uncovered*.
7. Peter Beaumont, 'Norwegian woman claims record time for climbing world's 14 highest peaks', *The Guardian* 27 July 2023. www.theguardian.com/world/2023/jul/27/kristin-harila-norwegian-claims-record-ascent-worlds-14-highest-mountains
8. Nadeem Badshah, 'Record-breaking mountaineer denies climbing over dying porter on K2', *The Guardian* 10 August 2023. www.theguardian.com/world/2023/aug/10/record-speed-mountaineer-denies-climbing-over-dying-sherpa-on-k2

Bibliography

This bibliography is divided into several sections. The main division, between primary material and secondary literature, should be self-explanatory. However, the primary material utilised throughout this book falls into what seem to me to be distinct categories. Under 'early modern' you will find the sources which formed the heart of this book: texts from the sixteenth, seventeenth and early eighteenth centuries which reveal the complexity of the European relationship with mountains before the age of mountaineering. As discussed in Chapters 4 and 5, the eighteenth and nineteenth centuries represented a transitional period both in terms of mountain attitudes and the development of the idea that 'they didn't like mountains back then', and texts relating to this transition can be found in their own subsection. 'Modern mountaineering writing' gathers together the writings of mountaineers from the earliest days of the Alpine Club up to the twenty-first century that have underpinned some of the later narratives of this book. The small subsection of 'news articles' bring us up to the present day.

Primary material

Early modern sources

Addison, Joseph, 'Ad Insignissimo Virum D. Tho. Burnettum, Sacræ Theoriæ Telluris Autorem', in *Musarum Anglicarum Analecta*, vol. 2 (1699).

Addison, Joseph, *Remarks on Several Parts of Italy, &c. in the Years 1701, 1702, 1703* (1705).
Addison, Joseph, 'The Pleasures of the Imagination', *The Spectator* no. 412, 23 June 1712.
Beaumont, John, *Considerations on a book, entituled the Theory of the Earth* (1693).
Bentley, Richard, *A Confutation of Atheism from the Origin and Frame of the World. The Third and Last Part ... Being the Eighth of the Lecture Founded by the Honourable Robert Boyle, Esquire* (1693).
James Brome, *Travels over England, Scotland and Wales* (1700).
Burnet, Thomas, *The Theory of the Earth: Containing an Account of the Original of the Earth, and of all the General Changes which it hath already undergone, or is to undergo, till the Consummation of all Things*, 2 vols. (1684–1690).
Burnet, Thomas, *An Answer to the Late Exceptions made by Mr Erasmus Warren against the Theory of the Earth* (1690).
Burnet, Thomas, *A Short Consideration of Mr Erasmus Warren's Defence of his Exceptions Against the Theory of the Earth* (1691).
Churchyard, Thomas, *The Worthines of Wales* (1587).
Coryate, Thomas, *Coryat's crudities hastily gobled up in five moneths trauells* (1611).
Croft, Herbert, *Some Animadversions Upon the Book Intituled the Theory of the Earth* (1685).
Davies, Robert (ed.), *The Life of Marmaduke Rawdon of York* (1863).
Defoe, Daniel, *A Tour thro' the Whole Island of Great Britain*, 3 vols. (1724–1727).
Drayton, Michael, *Poly-Olbion* (1612).
Evelyn, John, *The Diary of John Evelyn*, ed. E.S. de Beer, 6 vols. (1955).
Füssel, Stephan, *Theuerdank: The Epic of the Last Knight* (Taschen, 2018).
Gessner, Conrad, 'On the admiration of mountains' and 'A description of Mt. Fractus', trans. Dan Hooley in Sean Ireton and Caroline Schaumann (editors), *Mountains and the German Mind: Translations from Gessner to Messner, 1541–2009* (2020).
Higgins, John, *The Mirror for Magistrates* (1587).
[Kircher, Athanasius,] *The Vulcano's or, Burning and Fire-vomiting Mountains famous in the world* (1669).
Kirchner, Hermann, 'An oration on travel', translated in Thomas Coryate, *Coryat's Crudities* (1611).
Kirke, Thomas, *A modern account of Scotland being an exact description of the country, and a true character of the people and their manners* (1679).

Lithgow, William, *A most delectable, and true discourse, of an admired and painefull peregrination from Scotland, to the most famous kingdomes in Europe, Asia, and Affricke* (1614).
Lithgow, William, *The Pilgrimes Farewell, to his natiue countrey of Scotland* (1618).
Lithgow, William, *The Total Discourse, of the rare aduentures, and painefull peregrinations of long nineteene years trauayles* (1632).
Lovell, Archibald, *A Summary of Material Heads Which May be Enlarged and Improved into a Compleat Answer to Dr. Burnet's Theory of the Earth* (London, 1696).
Mackaile, Matthew, *Terræ Prodromus Theoricus* (1691).
Olearius, Adam, *The Voyages & Travels of the Ambassadors from the Duke of Holstein, to the Great Duke of Muscovy, and the King of Persia* (1662).
Petrarca, Francesco [Petrarch], letter to Father Dionighi da Borga San Sepolcro, April 26 1336, translated by Alan S. Weber in *Because It's There: A Celebration of Mountaineering, From 200 BC to Today* (2003).
Poole, Joshua, *The English Parnassus* (1657).
Pope Blount, Sir Thomas, *A Natural History: Containing Many not Common Observations: Extracted out of the Best Modern Writers* (1693).
Ray, John, *Select Remains of the Learned John Ray* (1760).
Ruchat, Abraham, *Les Délices de la Suisse* (1714), translated in William Windham, *An Account of the Glacieres or Ice Alps in Savoy* (1741).
Simler, Josias, 'On the Difficulties and Dangers of Alpine Journeys', translated by Alan S. Weber in *Because It's There: A Celebration of Mountaineering, From 200 BC to Today* (2003).
de Thévenot, Jean, *The Travels of Monsieur de Thévenot into the Levant* (1687).
Warren, Erasmus, *Geologia: A Discourse Concerning the Earth Before the Deluge* (1690).
Warren, Erasmus, *A Defence of the Discourse Concerning the Earth Before the Flood* (1691).
Warren, Erasmus, *Some Reflections upon the Short Consideration of the Defence of the Exceptions Against the Theory of the Earth* (1693).

Eighteenth- and nineteenth-century sources

Biese, Alfred, *The Development of the Feeling for Nature in the Middle Ages and Modern Times* (1905 English translation; original German edition 1888).
Gosse, Edmund, *Gray* (1882).

Rousseau, Jean-Jacques, *Julie, or the New Heloise: Letters of Two Lovers who Live in a Small Town at the Foot of the Alps* trans. Philip Stewart and Jean Vaché (*The Collected Writings of Rousseau*, vol. 6, 1997). First published in French in 1761.

Scheuchzer, Johann Jakob, *Ouresiphoites Helveticus* (1723).

Shairp, J.C., *On Poetic Interpretation of Nature* (1877).

Windham, William, *An Account of the Glacieres or Ice Alps in Savoy* (1741).

Wordsworth, William, 'Sonnet on the projected Kendal and Windermere railway', dated 12 October 1844, and letter to the editor of the Morning Post, dated 9 December 1844, reproduced in *The Prose Works of William Wordsworth*, ed. Rev. Alexander B. Grosart, 3 vols., vol. 2 *Aesthetical and Literary* (1876).

Modern mountaineering writing

Anonymous, 'Scheuchzer's Itinera Alpina', *Alpine Journal* vol. III (1867).

Anonymous, Review of *The Abode of Snow* by Andrew Wilson (1875), *Alpine Journal* vol. VII (1874–1876).

Cullinan, P. A. and G. Brammer, 'Australian Gauri Shankar (Tseringma) Expedition', *Himalayan Journal* 37 (1981).

Dent, C.T., *Mountaineering* (Badminton Library, London, 1892).

Dent, C.T., 'Mountaineering in the Old Style', *Alpine Journal,* vol. 11 (1882–1884).

Finch, G.I., *Climbing Mount Everest* (1930).

Farluund, Nils, 'The Tseringma Pilgrimage 1921 – an eco-philosophic "anti-expedition"', posted 14 August 2021, www.norgeshogfjellskole.no/the-tseringma-pilgrimage-1971-an-eco-philosophic-anti-expedition.

Freshfield, Douglas W., 'Placidus a Spesca and Early Mountaineering in the Bündner Oberland', *Alpine Journal*, vol. X (1880–1882).

Henderson, Bob, 'The story of the mountaineering anti-expedition of 1971', *Adventure Uncovered* published online 9 September 2021 https://adventureuncovered.com/stories/the-story-of-the-mountaineering-anti-expedition-of-1971.

Henderson, Bob, 'Rolwaling Legacy project – 1971–2021 (2023). https:/bobhenderson.ca/1971–2021-anti-expedition.

Irvine, Sandy, unpublished letters, held in the Sandy Irvine Archive, Merton College Oxford.

Longman, William, 'Modern Mountaineering and the History of the Alpine Club', appended to *Alpine Journal* vol. VIII (1878).

Mallory, George, unpublished letters from Everest 1921–1924, transcribed with kind permission from the Master and Fellows of Magdalene College, Cambridge.

Pollock, Frederick, 'The Library of the Alpine Club', *Alpine Journal* vol. XII (1884–1886).

Stephen, Leslie, *The Playground of Europe* (1871).

News articles

Badshah, Nadeem, 'Record-breaking mountaineer denies climbing over dying porter on K2', *The Guardian* 10 August 2023. www.theguardian.com/world/2023/aug/10/record-speed-mountaineer-denies-climbing-over-dying-sherpa-on-k2.

Beaumont, Peter 'Norwegian woman claims record time for climbing world's 14 highest peaks', *The Guardian* 27 July 2023. www.theguardian.com/world/2023/jul/27/kristin-harila-norwegian-claims-record-ascent-worlds-14-highest-mountains.

Francis Sanzaro, 'Keep Our Mountains Free. And Dangerous', *The New York Times*, 13 January 2018. www.nytimes.com/2018/01/13/opinion/sunday/keep-our-mountains-free-and-dangerous.html.

SECONDARY LITERATURE

Anand, Dibyesh, 'Western Colonial Representations of the Other: The Case of Exotica Tibet' in *New Political Science* vol. 29 (2007), pp.23–42.

Archer, Harriet, 'Mountains, identity and the legend of King Brennus in the early modern imaginary' in eds. Hollis and König, *Mountain Dialogues from Antiquity to Modernity* (2021), pp.197–214.

Bil, Albert, *The Shieling, 1600–1840: The Case of the Central Scottish Highlands* (1990).

Bosworth, Clifford Edmund, *An Intrepid Scot: William Lithgow of Lanark's Travels in the Ottoman Lands, North Africa and Central Europe, 1609–21* (2006).

Cajori, Florian, 'History of Determinations of the Heights of Mountains', *Isis*, vol. 12 no. 3 (1929), pp.482–514.

Calaresu, Melissa, 'Making and Eating Ice Cream in Naples: Rethinking Consumption in the Eighteenth Century' in *Past and Present* no. 220 (2013), pp.35–78.

Carlson, Allen, *Aesthetics and the Environment: The Appreciation of Nature, Art, and Architecture* (2000).
Clark, Kenneth, 'The Worship of Nature', *Civilisation: A Personal View by Kenneth Clark*, episode 11, BBC (1969).
Cleaver, Alan and Lesley Park, *The Corpse Roads of Cumbria: Walks along the county's ancient paths* (2020).
Colyer, Richard J., *The Welsh Cattle Drovers: Agriculture and the Welsh Cattle Trade before and during the Nineteenth Century* (1976).
Cosgrove, Denis, *Social Formation and Symbolic Landscape* (1954).
Cronon, William, 'The Trouble with Wilderness: or, Getting Back to the Wrong Nature', *Environmental History* vol. 1 no 1 (Jan. 1996), pp.7–28.
Debarbieux, Bernard and Gilles Rudaz, *The Mountain: A Political History from the Enlightenment to the Present*, trans. Jane Marie Todd (2015).
Duffy, Cian, *The Cambridge Companion to the Romantic Sublime* (2023).
Eisenstadt, S.N., 'Multiple Modernities', *Daedalus*, 129:1 (2000), pp.1–29.
Engel, Claire Éliane, *A History of Mountaineering in the Alps* (1950).
Findlen, Paula, *Athanasius Kircher: The Last Man who Knew Everything* (2004).
Fleming, Fergus, *Killing Dragons: The Conquest of the Alps* (2000).
Foucault, Michel, *The Order of Things: An archaeology of the human sciences* (Routledge, 2002, first published in English 1970, first published 1966).
Gilchrist, Paul, 'Gender and British Climbing Histories: Introduction', *Sport in History* 33:3 (2013).
Hahn, Dave, Eric R. Simonson and Jochen Hemmleb, *Detectives on Everest: The 2001 Mallory and Irvine Research Expedition* (2002).
Haldane, A.R.B., *The Drove Roads of Scotland* (first published 1952, reissued 2011).
Hansen, Peter H., *The Summits of Modern Man: Mountaineering After the Enlightenment* (Harvard University Press, 2013).
Heringham, Noah, *Romantic Rocks, Aesthetic Geology* (2004).
Hollis, Dawn and Jason König, ed., *Mountain Dialogues from Antiquity to Modernity* (2021).
Hollis, Dawn, 'The Contours of the North? British Mountains and Northern Peoples, 1600–1750' in eds. Dolly Jørgensen and Virginia Langum, *Visions of North in Premodern Europe* (2018), pp.223–42.
Hollis, Dawn, 'Mountain gloom and mountain glory: the genealogy of an idea', *ISLE: Interdisciplinary Studies in Literature and Environment*, 26:4 (2019), pp.1038–61.

Hollis, Dawn, 'Aesthetic Experience, Investigation and Classic Ground: Responses to Etna from the First Century CE to 1773', *The Journal of the Warburg and Courtauld Institutes*, 83:1 (2020), pp.299–325.

Hollis, Dawn, 'The "authority of the ancients"? Seventeenth-century natural philosophy and aesthetic responses to mountains', in eds Jason König and Dawn Hollis, *Mountain Dialogues from Antiquity to Modernity* (2021), pp.55–72.

Hollis, Dawn, 'The natural sublime in the seventeenth century' in ed. Cian Duffy, *The Cambridge Companion to the Romantic Sublime* (2023), pp.29–40.

Hoskins, W.G., *The Making of the English Landscape* (1955).

Isserman, Maurice and Stewart Weaver, *Fallen Giants: A History of Himalayan Mountaineering from the Age of Empire to the Age of Extremes* (2008).

Keenlyside, Francis, *Peaks and Pioneers: The Story of Mountaineering* (1975).

König, Jason, *The Folds of Olympus* (2022).

Larsson, Jesper, 'Labor division in an upland economy: workforce in a seventeenth-century transhumance system', in *The History of the Family* vol. 19 no. 3 (2014), pp.393–410.

McDonald, Donald, 'Lewis Shielings' in *Review of Scottish Culture* no. 1 (1984), pp.29–33.

MacDonald, Sadie, 'In Defense of Fanfiction: Authors as Fanfic Writers'. https://themindfulrambler.ca/2018/09/14/in-defense-of-fanfiction-authors-as-fanfic-writers.

MacFarlane, Robert, *Mountains of the Mind* (2003).

MacLean, Ian, 'Foucault's Renaissance Episteme Reassessed: An Aristotelian Counterblast', *Journal of the History of Ideas* (1998), vol. 59, no. 1, pp.149–66.

Mitchell, Ian R., *Scotland's Mountains Before the Mountaineers* (1998).

Mitchell, John, *The Shielings and Drove Ways of Loch Lomondside* (2000).

Neate, Jill, *Mountaineering Literature: A Bibliography of Material Published in English* (1986).

Nicolson, Marjorie Hope, *Mountain Gloom and Mountain Glory: The Development of the Aesthetics of the Infinite* (1959).

Porter, James I., *The Sublime in Antiquity* (2016).

Reuter, Benjamin and Jürg Schweizer, 'Avalanche triggering by sound: myth and truth', *International Snow Science Workshop* (Davos 2009), https://arc.lib.montana.edu/snow-science/item/252.

Roberts, Ian, Richard Carlton and Alan Rushworth, *Drove Roads of Northumberland* (2010).
Rossi, Paolo, *The Dark Abyss of Time: The History of the Earth and the History of Nations from Hooke to Vico*, trans. Lydia Cochrane (1984).
Salkeld, Audrey, *Last Climb: The Legendary Everest Expeditions of Mallory and Irvine* (1999).
Sauer, Carl, 'The Morphology of Landscape', first published 1925, reprinted in *Land and Life: A Selection from the Writings of Carl Otwin Sauer* ed., J. Leighley (1963), pp.321–3.
Scottish Archaeological Research Framework, 'Case Study: Transhumance and Shielings', *ScARF National Framework* (Modern Panel Report, September 2012). https://scarf.scot/national/scarf-modern-panel-report/modern-case-studies/case-study-transhumance-and-shielings.
Schellenberg, Betty A., 'Coterie Culture, the Print Trade, and the Emergence of the Lakes Tour, 1724–1787', *Eighteenth-Century Studies*, 44:2 (2011).
Summers, Julie, *Fearless on Everest: The Quest for Sandy Irvine* (2000).
Tatarkiewicz, Władysław, *A History of Six Ideas: An Essay in Aesthetics* (1980).
Thomas, Keith, *Man and the Natural World: Changing Attitudes in England, 1500–1800* (1983).
Toulmin, Stephen and Jane Goodfield, *The Discovery of Time* (1965).
Walton, Andrea, '"Scholar," "Lady," "Best Man in the English Department"? Recalling the Career of Marjorie Hope Nicolson', *History of Education Quarterly*, 40:2 (2000), p.176.
Weber, Alan S., *Because It's There: A Celebration of Mountaineering, from 200 BC to Today* (2003).
Williams, Gareth D., *Pietro Bembo on Etna: The Ascent of a Venetian Humanist* (2017).

Acknowledgements

The path to this book has been a long expedition during which I have received much support.

My academic research would not have got far at all without funding: thanks first and foremost then to the AHRC for funding my postgraduate research, to the Leverhulme for funding my postdoctoral work, and to the RSE for 'Covid re-boot' funding which enabled me to claw back some of the time lost to the pandemic childcare vacuum and complete a first draft of this book.

In the introduction, I shared that my path to the mountains of early modernity started in high school; so many thanks to the inspiring Amanda Richmond for starting it all with her loan of Everest books, and to Simon Gore and Andrew Soane for being the teachers whose example made me realise I wanted to become a historian. I am indebted to Lyndal Roper for sharing the early modern period with me, for suggesting that mountains were the most interesting thing I could choose to research, and for her unfailing generosity. I am hugely grateful to Alexandra Walsham, Bernhard Struck and Sarah Easterby-Smith for their supervision of my postgraduate research. Peter Hansen, as my PhD examiner, offered insightful feedback and has provided much-appreciated support to my career since.

Much of the writing of this book took place during my time as a post-doctoral researcher in the School of Classics at St Andrews. The School and its members provided a wonderful scholarly home to me for six years, and I am grateful to my colleagues both for their welcome and for (whether they intended to or not) expanding my horizons well beyond the early modern period. I have benefited, too, from an international community of mountain scholars. Special thanks to Zac Robinson, Stephen Slemon, and everyone else behind the 2015 and 2018 'Thinking Mountains' conferences, and much affection to Katherine Ledford and the wonderful members of the 2019 Appalachian mountain scholars' field trip.

Particular acknowledgement must go to Jason König, in whose company it has been a privilege and a pleasure to walk the paths of mountain research. I could not have asked for a more generous colleague and collaborator, or a more excellent scholar, to support me in the final few years of bringing this book into being. Anyone interested in the ways mountains were viewed in ancient Greek and Roman culture should be warmly encouraged to read his *Folds of Olympus*.

Emma Hart, Katie Ives, Jonathan Westaway and Tom Harrison offered their wisdom and advice when this book was still at the proposal stage. The manuscript itself has benefited enormously from the extra eyes and advice of Jason König, Andrew Szalay (the Suburban Mountaineer – look him up!), Eilidh Harris and her dad, Jim Pennel. Any errors, and indeed, errant commas, remain entirely my own.

After several years of searching for the right publisher I count myself enormously lucky that *Mountains Before Mountaineering* has found its home with The History Press. I am so grateful to Amy Rigg, my editor at The History Press, for putting her support behind this somewhat chimerical book (does it go in the history section in a bookshop? Sports? Adventure?), and for her patience with my endless questions about word length and image permissions. Enormous thanks must also go to Rebecca Newton, who as project editor guided me and my book to the final summit of publication, copy-editor Gaynor Haliday with her eagle eye for geographical details, and the typesetters and designers at The History Press for turning my text into a beautiful book.

Historians would not get very far without their sources. This book would not have been possible without the unnamed hands behind the digitisations of countless early modern books: thank you, whoever you are, for every page scanned. At the other end of the scale, 'Adam Olearius, my book' would not have been possible without the generosity and energy of Bill Zachs in establishing the Forbes Book Collecting Prize at the University of St Andrews. I am also grateful to the archives of Magdalene College, Cambridge, and Merton College, Oxford, for permission to reproduce the letters of George Mallory and Sandy Irvine, and to the archives of Smith College Massachusetts for providing me with long-distance access to materials relating to Marjorie Hope

Nicolson. Thanks also to Angus Vine for pointing me towards Marmaduke Rawdon and his antics at the summit of Mount Teide.

This book has been written at the sharp end of parenting a young family. Unconventional acknowledgements are therefore due to the wonderful staff of Rainbow Nursery, in whose care I have happily been able to leave my children whilst I have gone off to think and write about mountains. Love and thanks also to the fellow-parents whose company and support has kept me (relatively) sane whilst I grew my children and wrote this book: Dawn and Allan; Maarten and Sonja; Sarah and Sebastian; Cicely; Margaret; Eilidh; and Kimberley.

And of course, endless encomiums must go to my climbing partner in life, Kelsey Jackson Williams (who has never dropped me whilst belaying). In so very many ways this book would not exist without you. If you asked me the same question on the summit of Blencathra again, I would give the same answer. Finally, to V. and S.: I love you all the way to the mountains of the moon, and back.

Index

Note: mountains which have 'mount' as part of their full name (e.g. Mount Everest, Mont Blanc) can be found indexed under 'm'.

Addison, Joseph 169–70
alpages 62–3, 77, 92
alpenstock 87–8
Alpine Club, the 56, 97–8, 173, *185–8*, 193, 199
    see also Stephen, Leslie
Alpine Journal, the 97–9, 185–7
Alps, the 173
    attitudes towards the 7, 9, 169–70, 181–4, 186–8
cattle-herding in the 75–6
    see also transhumance
    early modern accounts of 24–8, 82–8, 90–2, 108
    see also Tarentaise Alps *and under* Burnet, Thomas
Anand, Dibyesh 69
Appalachians, the 65, 80
Atlas (myth of) 130–2
Atlas Maior 117–23
Aubry, Abraham 131
Augustine, Saint 55
avalanches
    early modern understanding of 83–5, 89
    on Everest 66–7

Balmat, Jacques 7, 184, 194
Baur, Johann Wilhelm 131
Beaumont, John 161, 163–4
Beerenberg, the 118–19
Bentley, Richard 166–7
Beş Barmaq 49
Blaeu, Joan 117–23
Boyle, Robert 166
Brennus (ancient British king) 108–9
Burckhardt, Jacob 53
Burke, Edmund 44, 164, 168
Burnet, Thomas 12, 147–9
    attitude to mountains 153–4, 179, 185, 206
    Nicolson, Marjorie Hope, on 189, 191
    responses to 154–67
    sublime, development of the 168–72

Theory of the Earth 150–2
    visit to the Alps 147, 151

Canary Islands *see* Mount Teide
Champaigne, Phillipe de 98–9
Christianity 144–6, 166
Churchyard, Thomas 103, 105
Clark, Kenneth 8–9
coffin stones *see* corpse roads
colonialism *see* imperialism
corpse roads 76
Coryate, Thomas *23–9*, 169, 200
Cosgrove, Dennis 102
crampons 88
crevasse
    avoidance 81, 87
    rescue 86–7
Croft, Herbert 159–61
Cronon, William 74
cultural geography 101–2
Cumberland and Westmorland 126, 171

Da Vinci, Leonardo 8
Defoe, Daniel 69–71, 169, 171, 177
Deluge, the 150–4, 170
Dennis, John 169
Dent, Clinton Thomas *173–4*, 199, 209
Derbyshire 169
dragons 7, 90–2, 184–5
Drayton, Michael 112, *123–6*

early modern (definition of) 10, 14, 16–18
    see also premodern *and* modernity
Englishman who Went up a Hill but Came down a Mountain, The 16

Farluund, Nils 203–4
Flood, the (Biblical) *see* Deluge, the
Foucault, Michel 161

Gabal Katrîne 32–6
Gatti, Oliviero 128
Gessner, Conrad *56–9*, 82, 181, 187
Goodfield, June 145
Gordon, Robert of Straloch 120–2
Gosse, Edmund 181
Gray, Thomas 179, 181
guides *see* mountain guides

Hansen, Peter 194–5, 197
Henderson, Bob 208
Highlands, Scottish 71, 73–4, 80, 94, 182
Hillary, Edmund 7, 11
Holy Land, the *32–41*, 119, 132, 136
Hoskins, W.G. 101
imperialism 33, 68
Irvine, Sandy 10–13

Jabal Musa 35–7
    see also Mount Sinai
Jan Mayen Island 118–19
Jericho 37
Jesus 37, 40, 98, 133, 136
Johnson, Samuel 182–3

Kant, Immanuel 44, 164
Kircher, Athanasius *139–41*, 184
Kirchner, Hermann 9

Lake District, the 10, 14, 124, 169, 171, 177–80, 205–6
Lancashire 124
'landscape', concept of 101–2
Lithgow, William 40–1, 43–5

Mackaile, Matthew 162–3
Mallory, George 10–11, 13, 33, 66–8, 148, 168, 198, 204
Matterhorn 102, 174, 209

Index

Maximilian I 104
Millet, Francisque 128–9
mining 48, 70–1, 78–9
Mirror for Magistrates, The 107–9
modernity 7, 194–8
    see also early modern
Momper, Joos de 128–9
Mont Blanc 7, 28, 61, 181, 184, 187, 194
Mont Cenis (pass) 27–8
Mont Ventoux 17, 53–5
Mormond Hill 120–3
Mount Catherine *see* Gabal Katrîne
Mount Etna 43–4, 58, 113, 140–1
Mount Everest 7, 83, 87, 200
    early attempts on 10–13, 33, 66–9
    as highest in the world 36, 51, 102
Mount Helicon 9, 58, 113
Mount Horeb 36
    see also Mount Sinai
Mount Ida 113–14
Mount Moses *see* Jabal Musa
Mount Olympus 9, 113, 200
Mount Parnassus 9, *43*, 58, *108*, 111, 113–14, 125
Mount Pilatus 56–9
Mount Pourri 62
Mount Quarantine *37–8, 40–1*, 132, 136
Mount Sinai 36, 39, 119–20, 132
Mount Tabor 132
Mount Teide (Peak of Tenerife) 43, *51–2*, 114, 163
Mount Vesuvius 58, 140–1
'mountain gloom' 8, 128, 168, 171, 176, 179, 189, 192
    see also Nicolson, Marjorie Hope
mountaineer, definition of 65
mountaineering 110, 173–4, 186–8, 194–5, 197, 199, 200, 203–5
mountain guides 25, 50, 81–2, 174, 181

mountains
    analogical ideas of 161–3
    in art 78–9, 126–36
    as beautiful 44, 143, 163, 166–7, 170, 181, 185
    definition of 15–16, 36–7
    and pilgrimage 32, 132
        see also Thévenot, Jean de
    in poetry 44, 103, 111–16, 123–5
    as sublime 44, 164, 167–8, 170–1, 185
    as useful 156, 163–4, 166–7, 182
    Virgin and Child depicted in 133–6
    women in the 18, 93–5, 207
Muir, John 74
muses, the 9, 43, 58, 111, 114
    see also Mount Parnassus

Næss, Arne 203–4
natural philosophy 36, 144–6, 166
Nicolson, Marjorie Hope 15, 109, 175–6, *189–92*
    see also 'mountain gloom'
Norgay, Tenzing 7, 11

d'Oggiono, Marco 133
Olearius, Adam 46–51
Orientalism 68–9
originality 109–10
Paccard, Michel-Gabriel 7, 184, 194
Paramount 102
Peak District, the 69–71, 169
Petrarca, Francesco (Petrarch) 9, 17, *53–6*, 59, 199
Pont, Timothy 120–2
Poole, Joshua 111–16
positivism 41
postmodernism 41–2
premodern (definition of) 17
Psyche (myth of) 130

Rawdon, Marmaduke 51–3
Ray, John 81
River Jordan 38, 40
Rocciamelone 28
Rossi, Paolo 147
Rousseau, Jean Jacques 180, 182, 183–4
Ruchat, Abraham 82–8, 182
Ruskin, John 176

Said, Edward 68–9
Sauer, Carl 101
Saussure, Horace Bénédict de 184, 194
Savery, Roelant 78, 128
Savoy 24, 82
Scheuchzer, Johann Jakob 90–2, 184–5, 186–7
science *see* natural philosophy
Setreng, Sigmund Kvaløy 203–4
Shairp, J.C. 180–1
Shakespeare, William 112, 180
shielings 77–8, 94–5, 207
Sicily *see* Mount Etna
Simler, Josias 75–6, 80–9
Skiddaw 10, 124–6, 205
snow trade, the 80
Snowdon 81
snowshoes 88
Steno, Nicolaus 146
Stephen, Leslie 109, 177, *181–5*, 188, 197
sublime, the 44, 164, 167–72
summit position, the *194–5*, 197, 199, 209

Tarentaise Alps, the 62–3
Tatarkiewicz, Władysław 165
Tempesta, Antonio 131
Tenerife *see* Mount Teide
Theuerdank 104–7
Thévenot, John de 32–9
Tibet 12–13, 66–9
Toblerone 102
Toulmin, Stephen 145
transhumance 77–8, 92–5
Typhon (mythological giant) 43

Valckenborch, Lucas van 79
Virgil 26, 58
volcanoes 44, 139–41, 153
    see also Mount Etna *and* Mount Vesuvius
Volga (river) 47–8
Vulcan (Greek god) 44

Warren, Erasmus 154–8
Western Elburz (mountain range) 50
Westmorland *see* Cumberland and Westmorland
Whymper, Edward 174, 209
wilderness, the idea of 74, 136
Windham, William 82
Wordsworth, William 7, *177–81*, 183, 188–9, 192, 197, 205, 208